The Language of Architecture and Civil Engineering

The Language of Architecture and Civil Engineering

By

Ana Mª Roldán Riejos, Joaquín Santiago López and Paloma Úbeda Mansilla

With a Preface by Alberto Campo Baeza

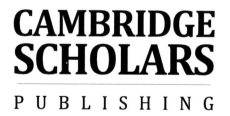

HUMBER LIBRARIES NORTH CAMPUS
205 Humber College Blvd
TORONTO, ON.　M9W 5L7

The Language of Architecture and Civil Engineering,
by Ana Mª Roldán Riejos, Joaquín Santiago López and Paloma Úbeda Mansilla

This book first published 2011

Cambridge Scholars Publishing

12 Back Chapman Street, Newcastle upon Tyne, NE6 2XX, UK

British Library Cataloguing in Publication Data
A catalogue record for this book is available from the British Library

Copyright © 2011 by Ana Mª Roldán Riejos, Joaquín Santiago López and Paloma Úbeda Mansilla

All rights for this book reserved. No part of this book may be reproduced, stored in a retrieval system, or transmitted, in any form or by any means, electronic, mechanical, photocopying, recording or otherwise, without the prior permission of the copyright owner.

ISBN (10): 1-4438-3167-0, ISBN (13): 978-1-4438-3167-3

To our students

"Language is inadequate to formulate the exact meaning and the rich variations of the realm of sensory experience"
—Moholy-Nagy L. 1947, *The New Vision*, Witenborn, New York

TABLE OF CONTENTS

List of Illustrations, Tables and Figures ix

Foreword xi
Dr. Jeannette Littlemore

Preface xiii
Alberto Campo-Baeza

Acknowledgments xvii

Abbreviations xix

Introduction 1

Chapter One 3
Message Organization in the Language of Architecture and Civil Engineering

Chapter Two 15
The Expression of Meaning in Technical Discourse

Chapter Three 25
Common Genres and Types of Texts

Chapter Four 37
The Use of Corpora Applied to Specific Language Purposes

Chapter Five 47
Textual Strategies: Language and Use

Chapter Six 57
Metaphor in the Domains of Architecture and Civil Engineering

Bibliography 75

About the Authors 79

LIST OF ILLUSTRATIONS, TABLES AND FIGURES

Ex. 1-1 Sketch Guggenheim Museum Bilbao (Spain) by Frank Ghery 10
Ex. 1-2 Sketch Puente de Ourense (Spain) by Alvaro Varela 10
Ex. 2-1 Pyramids 16
Ex. 2-2 Taj Mahal 16
Ex. 3-1 Millennium Bridge 32
Ex. 4-1 Concordance lines for *structure* 40
Ex. 4-2 Sample context for *structure* 41
Ex. 4-3 Total number of word tokens and different word types 42
Ex. 4-4 Sample of clusters for *structure* 43
Ex. 6-1 Metaphor Mapping 59
Ex. 6-2 Visual metaphor provided by the architect: Teeth 62
Ex. 6-3. Visual metaphor provided by the architect: Brain 62
Ex. 6-4 Metonymy Mapping 65
Ex. 6-5 Visual metaphor provided by the architect: Cubes 67
Ex. 6-6 Visual metaphor provided by the architect: Ship 69

Table 1-1 4
Table 2-1. Components of academic and technical communication 17
Table 2-2. Polysemy in "strain" 21
Table 6-1. Steps used by Steen (1999: 57) and the Pragglejaz group (2007)
    to identify metaphors 64

Figure 2-1 Constraints on interpretation 19
Figure 2-2 Contextual disambiguation of "bleeding" 20

FOREWORD

In this volume, Ana M. Roldán-Riejos, Paloma Úbeda and Joaquín Santiago provide a comprehensive and in-depth overview of the language of architecture and civil engineering, by presenting several analysis tools and approaches. It draws on an impressive range of disciplines such as genre theory, register analysis, corpus linguistics and metaphor theory. The result is an inspiring volume which will be of interest to architects, civil engineers, linguists and language teachers. They first chapter begins by introducing the main types of architecture and civil engineering texts and genres and discusses the typical rhetorical functions and characteristics that may be found therein. It adopts a genre-based approach to the discipline and dicusses how meaning develops and is expressed in scientific-technical discourse, rightly pointing out the extent to which metaphor is used to develop and convey abstract concepts within these disciplines. Apart from linguistic matters, the multimodal nature of metaphor and its importance in academic discourse are considered. Thus providing useful information for practitioners and teachers of English for Specific Purposes on how to write appropriately in the different genres that are involved in the language of architecture and civil engineering. A particularly novel feature of the book is its chapter on corpus compilation and the use of freely-available corpus linguistic software (Antconc) which provides learners with the tools to take over their own learning and to tailor their linguistic input to their specific needs. This chapter, which is set out as a kind of 'how to' guide, would also be very useful for university teachers who are engaged both in teaching and researching in English for Specific Purposes. This book also offers useful suggestions for small action research projects that final year students might engage in with the support of their tutors. The book uses findings from functional grammar research to make suggestions to writers wishing to improve their rhetorical techniques. Areas that are covered include thematic progression, the use of adverbs and modal verbs, sentence arrangement, hedging and appropriate use of the passive voice. The discussion of metaphor and metonymy is particularly insightful as it draws the student's attention to both conceptual and linguistic metaphors and metonymies and looks at how they work together across stretches of discourse in order to create and convey meaning.

All in all, this book constitutes and excellent guide to teachers, practitioners and students alike, working in the fields of architecture and civil engineering. The book is clearly written with concise summaries and suggestions for further reading at the end of each chapter. It is expected to be a useful resource for teachers of English to students working in these disciplines as well as serving as an excellent self-study book for people already working in the field who need to communicate in English. The authors are experienced ESP teachers and researchers and it is easy to see that this book is the result of many years working in the field.

—Dr. Jeannette Littlemore
Senior Lecturer, University of Birmingham,
United Kingdom

PREFACE

(Spanish version)

WORDS AS BIRDS
Al igual que las palabras en alas de la Poesía son capaces de volar al cielo como los pájaros, las palabras montadas en la lengua de Shakespeare son capaces de llegar hasta el más recóndito lugar de este mundo, y más aún con las nuevas tecnologías.
Pues este libro, lleno de las palabras clave de Arquitectura e Ingeniería e Informática, en inglés, será un instrumento eficaz para escribir y hablar con precisión en inglés sobre estos temas específicos.
Aquí se dan instrumentos a los arquitectos y a los ingenieros, a los profesores y a los estudiantes, para comerse el mundo... en inglés.
Estas palabras inglesas, precisas, harán posible que la comunicación se haga con precisión. La precisión exigible en la transmisión de los conocimientos y en la comunicación de la Arquitectura y de la Ingeniería. Tan universal es la lengua inglesa.

CERVANTES
Esta universalidad del inglés la entendió muy bien Cervantes, que parecía tan serio, cuando nada más terminar de escribir D. Quijote de la Mancha, encargo, él mismo, la traducción al inglés a un tal Shelton en 1612. Curiosamente el mismo año en que Homero era traducido también al inglés por vez primera por Chapman.
Tan lejos llegaron las palabras del Quijote, que, Jefferson, el tercer presidente de los Estados Unidos que era arquitecto, reñía a su hija María porque "no estaba leyendo D. Quixote de la Mancha" como él le había indicado.

COMUNICACIÓN
Y si el dominio de la lengua inglesa es importante, todavía lo es más cuando se usa el ordenador donde el inglés es imprescindible.
Cuando en mi ordenador aparecen emails en chino o en japonés, o en lenguas que yo no entiendo, los borro inmediatamente. No así con las muchas cartas que vienen en inglés, como lo más natural del mundo. Además hay palabras en castellano que parece no serían fáciles de traducir, como infalible e inefable, y en inglés son inffalible and ineffable. Tan sencillo como eso. Claro que brújula, plomada y nivel, los tres

instrumentos imprescindibles para un arquitecto, son "compass", "pluma" y "spirit level".
  Cuando se tradujo mi libro "La idea construída" al inglés, el primer borrador decía "The constructed idea". Tuve que insistir en que mi intención era transmitir "The built idea", la idea construida, como la idea de un arquitecto que se materializa en un edificio construido. Este "The built idea" que lleva ya más de 20 ediciones en Castellano, 4 en portugués, 1 en francés y 1 en inglés editada por japoneses, va a ser editada por una editorial china en inglés. Con una gran tirada para, inmediatamente ser traducida y difundida en chino.

DOCENCIA
Hay situaciones en las que es necesario para los arquitectos y los ingenieros y los estudiantes de estas áreas, tener un inglés perfecto. El mejor posible, no sólo en la gramática, sino en los términos técnicos. Cuando se ejerce la docencia o se dan conferencias en el extranjero. Yo he dado mis conferencias en Japón o en Noruega o en Alemania en inglés. Y mi docencia en Penn no podría haberla dado en otra lengua que no fuera el inglés.
  Aún recuerdo las risas de los asistentes cuando en mi primera conferencia en Italiano, en Italia, hablé de una "coperta a due aque" que significa una "manta a dos aguas" que no significa nada. Debí decir "tetto a due falde" que es, en correcto italiano, la traducción de la "cubierta a dos aguas". En estas materias hay que hablar con palabras exactas para transmitir bien las ideas. Y también para aprender, para estudiar.

FINAL
Que alguien elabore un libro que ponga al día los términos técnicos necesarios para los estudiantes, los arquitectos y los ingenieros, es un regalo. Este es un libro que debería haber aparecido hace tiempo en el mercado y que, ahora, hay que agradecer a sus autores. Un libro que nos da alas para poder volar con las nuevas tecnologías y que no debe faltar en ninguna biblioteca universitaria.

(English version)

WORDS AS BIRDS
As words, like birds, can swirl up in the air on the wings of Poetry, so can they ride on the back of Shakespeare´s language to reach the most secluded places in this world, especially with the help of new technologies.

Thus, this book full of English words key in Architecture, Engineering and Computing shall be an efficient tool to write and speak accurately in English about these specific fields. The way is open for architects and engineers, professors and students to take the world by storm.

With precise English words communication can therefore be precise too. Precision is required to transmit knowledge and to communicate Architecture and Engineering. The English language is that universal.

CERVANTES
The English language's universality was well understood by Cervantes, so severe in appearance, when upon completing *Don Quixote of La Mancha*, he commanded himself a translation into English to Shelton in 1612. Curiously enough, that very same year Homer was also translated into English for the first time by Chapman.

Don Quixote's words reached such a scope that Jefferson, the third US President, who happened to be an architect, used to tell his daughter off because "she wouldn't read her *Don Quixote of La Mancha*" like he had prescribed her.

COMMUNICATION
Not only is the command of English important, it is paramount when using computers for which English is essential.

As soon as I get an email in Chinese or Japanese or in another language I can't understand, I delete it right away. That is not the case of the many letters written in English for they are naturally taken in. Moreover, there exist words in Spanish that might seem hard to translate such as *"infalible"* or *"inefable"* and which simply translate into "infallible" and "ineffable". However, *"compass"*, *"pluma"* and *"spirit"*

When my book *"La idea construida"* was translated into English, the first draft was titled "The constructed idea". I had underlined that my intention was to transmit "the built idea", not unlike an architect's idea coming true in the shape of a finished building. "The Built Idea" has over 20 editions in Spanish, 4 in Portuguese, 1 in French, 1 in English done by

Japanese publishers and it will be edited by a Chinese publisher in English to be translated and distributed in Chinese.

LECTURING
In certain situations, architects, engineers and students in this field need to have a perfect command of English. The best they can, not only in terms of grammar, but also regarding technical terms. This is especially true when lecturing or giving conferences abroad. I have given lectures in English in Japan, Norway and Germany. My lectures in Penn could have been given in no language other than English.

I still remember the attendants of my first conference in Italian, in Italy, sniggering when I mentioned a "coperta a due aque" which means a "double-watered blanket" and has no meaning whatsoever (I should have said "tetto a due falde" which translates correctly to "gable roof"). In fields such as these, the right words are necessary to transmit ideas successfully. This is also true when learning or studying.

FINAL
Having someone compiling in one book and updating all the specific technical terms needed by students, architects and engineers is a gift. This book should have been launched into the market long ago and makes its authors worthy of our recognition. This book gives us wings to fly again on the wings of new technologies and shouldn't be missing in any university library.

—Alberto Campo-Baeza

Acknowledgments

It is a pleasure for us to thank those who made this project book possible, such as our students from Language of Architecture and Engineering, subject in the Technical University of Madrid (Spain), for motivating us with their feedback for putting together this project. We highly appreciate the effort of those who supported us during the completion of the book. Also we want to thank the collaboration of Dr. Jeannette Littlemore for her generous comments in the *Foreword*. Finally, we are heartily grateful to the well-known Spanish architect Alberto Campo-Baeza for kindly accepting to write the *Preface* of this book.

ABBREVIATIONS

ACE	Architecture and Civil Engineering
CL	Corpus Linguistics
ESP	English for Specific Purposes
LACE	Language of Architecture and Civil Engineering
LSP	Languages for Specific Purposes
RCC	Reinforced Cement Concrete

INTRODUCTION

This book has been written with three concentric audiences in mind, the most central being linguists and Languages for Specific Purposes (LSP) practitioners. The next ring includes those architects and civil engineers willing to deeply understand and to be aware of the use of their professional language from a comprehensive approach that takes into account cognitive, social and pedagogical factors. This would enable them to adapt it to their needs and to exploit it conveniently. The outer ring comprises architecture and civil engineering students who partake in a wide scope content and conceptual core, and who may use English as an academic and professional vehicle of communication now or in future work scenarios. According to the European Commission, language learning is a lifelong skill that needs practical tools of which this book could be a useful example.

The structure of the book has been designed to be read and used independently of its content outline by anyone interested in these areas. The book is divided into six chapters; each of them being the result of previous research carried out from various interdisciplinary angles, and has been addressed from the language and communication approach. Another innovative feature of this work is that new approaches to language and linguistic studies have been incorporated and addressed. For example, recent findings on Corpus Linguistics and Cognitive Linguistics, such as metaphor, meaning interpretation and lexical collocation analysis have been highlighted. Text studies and new electronic genres such as academic and professional e-mails have also been considered. The arrangement of the chapters follows a similar format; each one based on theoretical background contents, and has been written in a clear and accessible style. The chapters also contain a practical section with examples and follow-up tasks inspired or taken from authentic engineering and architectural materials such as journals (including electronic ones), essays, research textbooks, etc.

In sum, this volume has been designed to cover and highlight, both for the novice and the expert, the main features that shape and define the type of communication and main genres that architects and civil engineers use and deal with throughout their training and careers.

Chapter One

Message Organization in the Language of Architecture and Civil Engineering

This chapter deals with:

✓ The main types of architecture and civil engineering texts and genres.
✓ The typical sections they may contain.
✓ The rhetorical functions and characteristics that may be found in LACE.

Do you know that...?

The analysis of any piece of written information can be made in different ways depending on our aim and expectations about the text. Traditionally, two approaches have been used to disclose the inner texture of a text, namely the "top-down" approach and the "bottom-up" approach. The first one consists of exploring the overall organization of the text, e.g. focusing on major graphic parts, textual functions and structural conventions (macro-level). Thus, we get familiar with the different structures of a journalistic article, a doctoral thesis, an engineering report, an abstract, or an instruction manual. By contrast, the bottom-up approach is concerned with the analysis of the different components of the text, such as the choice of words or the syntactic arrangement (micro-level). Obviously, both levels of analysis can be used to study any ACE text.

The notion of *genre* basically refers to the macro-level organization of discourse. This applies to written or spoken genres, though the present work will rather focus on the former ones. It is through the study of genres that we are able to see "the big picture" of any piece of discourse and to make sense of the organization of messages. Here is a list of some of the most common LACE academic and professional *genres*, some of them are dealt with in subsequent chapters of this book.

❖ Abstract	❖ Report
❖ Research work	❖ Thesis
❖ Professional e-mails	❖ Business Letters
❖ Journal Articles	❖ Resume/CV
❖ Specifications	❖ Instructions

Table 1-1.

According to John Swales, "*The principal criterial feature that turns a collection of communicative events into a genre is some shared set of communicative purposes*" (Swales, 1990: 46). Most genres use conventions related to communicative purposes, for example, a business letter may start by referring to a published job position, because its purpose would be to apply for a specific job and its requirements. These conventions are shared by the addressee (the company offering a job) and the sender (the person applying). In an argument essay, the writer would emphasise their thesis since the aim is making a point. Thus, the specific social goals become main foci when genre is discussed. It also implies that before writing, the context of a situation should be considered and analyzed in order to anticipate the linguistic features that are required. Some examples of genres may be more prototypical than others, depending on their structure, formal aspects, etc. however, for genre membership some basic resemblance should remain, which according to Swales would be its communicative purpose (Ibid.49-50).

I. An example of academic genre: abstracts

An abstract is a summary consisting of a limited number of words referred to a completed research work. If done well, it makes the reader want to learn more about that research. It usually precedes research articles published in specialised journals and its length in LACE may range from 100-250 words.

These are some basic components of an abstract in any discipline:

1) Stating the problem/motivation: Why are we concerned about the problem? What practical, scientific, theoretical or artistic gap is that research trying to fill? E.g. To study the location of a new building in a historical town central area.

2) Methods/procedure/approach: What did we actually do to get these results? E.g. we analysed 3 bridges, completed a series of short projects, interviewed 17 architects)
3) Results/findings/product: As a result of completing the above procedure, what have we learnt/invented/created? E.g. we have shown the advantages of a type of concrete or material.
4) Conclusion/implications: What are the larger implications of our findings, especially for the problem/gap identified in step 1? E.g. we provide the environmental implications of inserting a dam in a specific site.

When the abstract forms part of a journal article or paper to be published, it is common to add four or five *keywords* at the end which make reference to the main ideas expressed in the text.

Example of ACE abstract:

In this paper the Italian CNR-GNDT vulnerability index for masonry buildings was modified to apply in confined masonry buildings and to obtain a reasonable relationship with the wall density per unit floor index. With this purpose, a sample of twenty-four confined masonry buildings with three and four storeys built during the last twenty-five years for social housing programs was used. A relationship has also been obtained between the value of the proposal index and the damage observed in the March 1985 Central Chile subduction earthquake (M_s=7.8).

Keywords: *Seismic vulnerability; Confined masonry buildings; Relationship vulnerability index–damage.*

Source: Kenneth, A., Gent, F., Gian, M., Giuliano, M., Maximiliano, A., Astroza, I., Roberto, E. (2008) 'A seismic vulnerability index for confined masonry shears wall buildings and relationship with the damage', *Engineering Structures*: 2605-2612.

II. An example of professional-academic genre: Reports

There are many types of academic or professional reports. In LACE you may need to write a report on a particular work to present to your boss or maybe you have to write a report for a team of engineers or for a mixed audience that includes other professions of non-experts in the field. When preparing a report, these are important factors to bear in mind. Typically, the structure of a report would cover the following:

1. **Introduction**: It provides a general overview of the work. It usually includes a literature review on the project to be developed, or specifies background reading on the project. The introduction also shows previous preparation before commencing the work. This section will enable you to explain what makes your project interesting and distinct from previous works as well. The introduction can be finished with the aims of the project.
2. **Methodology**. The procedure followed may now be written, explaining with as much detail as required for the reader to understand everything that was done: work execution, problems encountered with precise data and numerical figures. It is usually written in past tense and using passive voice. Appendices are sometimes generated during the writing of the core of the report.
3. **Results/Discussion**. In these two sections, both textual and visual information are included: graphs, diagrams, maps, that may contain information to explain work development, expected or unexpected outcomes and on how possible problems were solved.
4. **Conclusions**. The conclusion will highlight the most relevant parts of the work by stating concisely major aspects of the results and discussion. The conclusions should not include new material. This section basically condenses the content of earlier sections. Ideally, it should be written bearing in mind readers who may wish to become familiar with your work without knowing the fine details.
5. **Appendices.** Finally, a summary may be written. This is not new material either, and should include key points from the introduction and the conclusions.

Furthermore in the field of architecture report genres, we can find a sub-genre design report where we can find further reports such as feasibility report, sustainability report, conservation area report, listed building consent report, traffic report, ecology report etc.

Basically these sub-genres consist of a "design statement", page section and previous named reports depending on the purpose of the project submission. However the most frequent used report genre for students in an academic context is the descriptive report of a project.

To see examples of ACE reports, you can click the following link:
<http://www.civil.usyd.edu.au/publications/2004rreps.shtml#r835>

Each academic and professional genre may contain rhetorical functions according to the text type and to the communicative intentions of the author, as we can see below:

III. Main rhetorical functions of ACE texts and the genres they may belong to

a) **Descriptive**

If we have to describe a *building site* where a structure will be built, then the main function of the text will be descriptive. Therefore we may use verbs of 'non-change' (e.g. *to be, to stand, lie, rest,* etc.) and adverbs of place *(away, up, in, down, underground, outdoors, outside)*. Technical descriptions tend to be neutral, exact and impersonal. Depending on what we are describing, descriptions can be:

i) Mechanism/Location description

It may explain the arrangement and shape of an object in space or the ins and outs of a particular location where a structure will be built. Typically, the parts of mechanism description answer the following questions:

- What is it?
- What is its function?
- What does it look like?
- How does it work?
- What are its main parts?

Thus, the description of a building location would answer these questions:

- Where is it? What will its function be?
- What is its orientation and shape?
- What type of area is it (industrial, residential....)
- What problems may the building entail?
- What type of planning regulations may apply?

An example of this type of description may be describing what a crane consists of. Such a description may involve movement. Complex motions are better handled with process description.

ii) Process description

It explains the arrangement of a sequence in sequential order. It is similar to mechanism description, except that the part-by-part becomes step by step, answering questions such as:

- What is it?
- What is its function?
- Where and when does it take place?
- Who or what performs it?
- How does it work?
- What are its main steps?

For instance, we may describe how a bulldozer operates or how a work in a building site progresses.

Descriptions can be used in various technical genres, for instance research journal articles, they could be a part of a report, or used in technical manuals and textbooks.

b) Narrative

If we need to write a report about the steps needed when building a *dam*, for example, then we are talking about a narrative text. Narrative texts types deal mainly with changes in time, i.e. with actions and events. Typical text type markers would be verbs that denote 'change' as well as expressions of time (time-sequence signals); likewise adverbs of place could also be used. Genre type: the narrative function is typically used in technical reports.

c) Instructive

If we are going to give precise information about how to operate with a *paving machine* in highway construction, or about enforced *environmental regulations*, then instructive texts will be used. In this type of text either the imperative mood or the passive voice with modal verbs *must or should* are usually used. Genre type: Technical manuals, leaflets with technical specifications.

d) **Expository**

If we are explaining the various stages of *concrete* placing or providing information about ideas to design a cable-stayed bridge, then we are probably using an expository text. Usually, the main ideas are explained by means of supporting details, by giving examples and specific data. Typical verbs for the identification and explanation of objects and ideas are: *to refer to, be defined, be called, consist of, contain* etc. If we have to establish a relation to previously mentioned facts and ideas, we may use expressions such as: *namely, incidentally, for example, in other words*, etc. A similarity to preceding phenomena can be expressed by: *similarly, also, too*; additional information can be indicated by expressions like: *in addition, above all, moreover*, etc. This function is typically used in genres such as expository essays, summaries, also being frequently used in textbooks.

e) **Argumentative**

When writing to give reasons for a new theory, idea, or viewpoint that we intend to show to other engineers or architects, we are probably using argumentative texts. Reasons for or against a number of topic are put forward. The ultimate goal is usually to win the reader/audience round to the author's side. This is a leading contrastive text structure, and expressions like *but, by contrast, however, yet, still, in any case, so*, etc. work as its linguistic signals. However, the basis of any argumentative text is provided by expository passages, by the explanation of facts, concepts, developments or processes. While comments tend to be subjective in character, scientific arguments seek to be objective. The argumentative function may be used in journal articles, abstracts, or parts of a report.

Let us focus now on some major characteristics of architecture and civil engineering texts:

IV. Typical features of ACE texts

a) **Use of text and visual information**

ACE texts often include lots of visuals that help to illustrate and cast light on the meaning expressed. Visual display is part of the training that architects and engineers receive and therefore it is a feature in designs, projects, and a basic part in their communication. Visuals may consist of charts, diagrams, pictures, drawings, graphs, etc. Visual information frequently includes explanatory texts to guide about the type of visual information exposed, or stating why that information may be relevant. In this sense, explanatory texts reinforce visuals. According to Trimble, visuals "provide us with detail difficult to explain in words" (Trimble, 1985: 103). On the other hand, numerical data, graphs and sketches form part of architects and engineers training, thus it is important to give clear clues to interpret them. Furthermore, visuals may help to exemplify instructions, descriptions, reports, laboratory experiments or any other type of technical discourse.

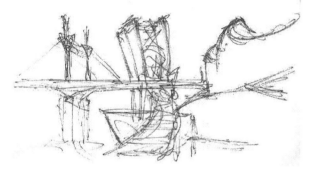

Ex. 1-1 Sketch Guggenheim Museum Bilbao (Spain) by Frank Ghery

For instances, in architecture and engineering contexts drawing are produced to provide a better idea either about the project itself or a about detail. A sketch can be nowadays entirely computer-generated.

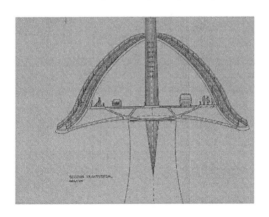

Ex. 1-2 Sketch Puente de Ourense (Spain) by Alvaro Varela

b) LACE characteristic style

If we assume that engineers and architects discourse[1] is mainly concerned with facts, the use of aesthetic techniques is considered to be out of the question. Indeed, ACE discourse seems to favour a straightforward and simple style usually loaded with technical terms and an impersonal voice. Yet, as we will show in this volume, the occurrence of rhetorical patterns that help to modulate ACE communication is by no means atypical. Interestingly, ACE language involves the use of strategic features as we will see in successive chapters. Let us consider the following case:

The distinction between the initiation life and propagation life is important. At low strain amplitudes up to 90% of the life may be taken up with initiation, while at high amplitudes the majority of the fatigue life may be spent propagating a crack. Fracture mechanics approaches are used to estimate the propagation life.

The strengthening of reinforced concrete members by externally bonded steel plates has become increasingly common because of its cost-effectiveness and versatility. These days, the cracks caused due to the heavy loading have made the structure less durable. Hence more recent activity has been directed in long term durability of the adhesive and its interface bonding with the steel plate and the concrete.

From: SASIKUMAR, P.A. (2006) *Interfacial Fracture Mechanics: Patch Repaired System in Beam.* International Institute of Information Technology Hyderabad (Deemed University).

Available at:
<http://EnrD2svnxV8J:sites.google.com/site/arjunsas>

In the first paragraph, we can see how the author presents a point through a twofold comparison (*At low strain..... while at high...*) and uses `may´ to give an example supporting this assertion. In the second paragraph, the author manages to keep an impersonal voice through the use of stylistic means, for example, subjects with no agentive role or by the passive voice. The author manages to do this despite inserting a

---

[1] We refer here to communication in English language. Stylistic features and conventions used in academic and professional Spanish may differ.

criticism on previous work undertaken in the field. This represents an elaborate way to communicate a given position without posing a threat to a possible technical audience.

V. Summing-up

In this unit, we have seen most representative genres for the students of architecture and engineering could use in academic context. We have also explained the main features of these texts and the typical sections they might contain. Finally, in this unit we have outlined the role of the basic rhetorical functions and characteristics paying special attention to a common profile focus on the language of architecture and civil engineer.

VI. References

CARTER, D. (1994). *Introducing Applied Linguistics: an A-Z Guide.* London: Penguin.
HERBERT, A. J. (1969). *The Structure of Technical English.* Harlow: Longman.
KENNEDY, C. & R. BOLITHO (1984). *English for Specific Purposes.* London: MacMillan.
HYLAND, K. (2004). *Genre and Second Language Writing.* Ann Arbor, MI: University of Michigan Press.
SWALES, J.M. (1990). *Genre Analysis: English in Academic and Research Settings.* Cambridge: Cambridge University Press.
SWALES, J.M. and C.B. Feak. (2010). "From text to task: Putting research on abstracts to work". In Ruiz-Garrido, M; Palmer-Silveira, J.C. and Fortanet-Gómez, I.: *English for Professional and Academic Purposes.* Amsterdam-New York: Rodopi.

Follow- up practice unit 1

1. **Read the 2 examples below and indicate:**
   a) **Main Function/s**
   b) **Type of Text**
   c) **The parts they include**

Example A

Concrete is a synthetic construction material made by mixing cement, fine aggregate (usually sand), coarse aggregate (usually gravel or crushed stone), and water in the proper proportions. The product is not concrete unless all four of these ingredients are present. CONSTITUENTS OF CONCRETE The fine and coarse aggregates in a concrete mix are the inert, or inactive, ingredients. Cement and water are the active ingredients. The inert ingredients and the cement are first thoroughly mixed together. As soon as the water is added, a chemical reaction begins between the water and the cement. The reaction, called hydration, causes the concrete to harden. This is an important point. The hardening process occurs through hydration of the cement by the water, not by drying out of the mix. Instead of being dried out, concrete must be kept as moist as possible during the initial hydration process. Drying out causes a drop in water content below that required for satisfactory hydration of the cement. The fact that the hardening process does not result from drying out is clearly shown by the fact that concrete hardens just as well underwater as it does in air.

Example B

Earthquake-induced structural pounding may result in considerable damage or even collapse of colliding structures during severe ground motions. The aim of the paper is to conduct a detailed investigation on pounding-involved response of two equal height buildings with substantially different dynamic properties, paying a special attention to modelling the non-linear effects taking place during impact as well as observed in the structural behaviour as the result of ground motion excitation. The three-dimensional non-linear response analyses as well as the parametric study have been conducted for earthquake-induced pounding of structures modelled as elastoplastic multi-degree-of-freedom lumped mass systems. The results of the response analysis show that collisions have a significant influence on the behaviour of the lighter and more flexible building causing substantial amplification of the response and leading to considerable permanent deformation due to yielding. On the other hand, it has been found that the behaviour of the heavier and stiffer building is nearly unaffected by collisions between structures. The parametric investigation has led to the conclusion that the peak displacement of the lighter and more flexible building is very sensitive to a change of structural parameters considered (gap size, storey mass, structural stiffness and yield strength), whereas the response of the heavier and stiffer structure can be influenced only negligibly.

Keywords: Structural pounding; Earthquakes; Buildings

2. Find 2 examples of ACE abstracts and indicate whether they follow the "standard" structure. If not, indicate the sections as well.

3. Look for examples of the interrelation between text and visuals in the field of architecture and civil engineering, and:

3.1. State what type of discourse the visual is illustrating (instruction, report, etc.).
3.2. Give your opinion about its contribution to understanding the text content.

See: www.LACE.aq.upm.es for sample exercises from this unit.

CHAPTER TWO

THE EXPRESSION OF MEANING IN TECHNICAL DISCOURSE

In this chapter you will learn about:

✓ How meaning develops and can be expressed in scientific-technical discourse
✓ The importance of context in communication
✓ Factors that may affect meaning de-codification

Introduction

We can express meaning in many ways, e.g. in art, in language, and also in general social interaction (social events, rituals, gestures, body language etc.). Human beings are capable of producing and interpreting meaning, but meaning does not exist *per se*. Instead, the participants in communicative acts contribute in the creation, development and interpretation of 'meaning'. Consider for instance the ancient Pyramids: they were primarily meant to keep dead Pharaohs undisturbed in their graves for the centuries to come. Egyptians believed in life after death. Nowadays the pyramids are considered a symbol of Egypt. Or take the Taj Mahal: a fascinating Indian palace interpreted by many as representing love. These two next examples illustrate the expression of meaning in technology and architecture.

16 Chapter Two

Ex. 2-1 Pyramids

Ex. 2-2 Taj Mahal

I. Meaning and communication

Meaning is understood through our interaction with the world. Starting from bodily sensor-motor experiences, infants start to recognize most concepts, hence we are able to understand and learn by using perceptual, emotional and mental operations. For example, we acquire a direct experience of vertical direction, i.e. *up* and *down*, as well as inclusion/ exclusion, i.e. *in* and *out*. This represents a cognitive experientialist approach (Lakoff and Johnson, 1980) different from rationalist theories in which reason is the only source of knowledge interpretation and creation. As Lakoff & Nuñez (2000) have shown, most mathematical reasoning arises in bodily experience and depends on bodily configuration, rather than on disembodied mind operations.

Accordingly, meaning derives from cultural experience and is stored in our brain so that new concepts "make sense" or "match" mental categories—aka spaces—previously configured in our brain. Therefore, our cognitive abilities make communication possible through the interpretation and expression of meaning. It is difficult to communicate without meaning.

By communication we understand a combination of interactive grammatical, social, pragmatic and cognitive factors. These elements work together; therefore, it is not easy to separate them except for theoretical purposes. This definition can be applied to academic, scientific and technical communication. In addition, every form of communication occurs in a specific context; therefore context is a major element in producing and interpreting meaning.

II. Context and communication

Context is one of the main components of communication, though there also other elements to consider, as we can see in the table below:

The type of audience	The type of discipline
The speaker/writer	The type of genre
The context	

Table 2-1. Components of academic and technical communication

Context is not only considered linguistically, there are extra linguistic factors that may influence communication, together with psychological, social and situational conditions. This includes how context is conceptualised, or "internalised" by the speaker and the hearer. Hence, context is also viewed as a mental phenomenon. In a conference scenario, for example, the participants need to know the code of conduct implied by that context which may restrain their behaviour. For example, they are not expected to sing or dance during a presentation; instead, they are supposed to listen attentively to speakers and to ask questions when appropriate.

In academic communication, both spoken and written forms of language may co-exist and, consequently, the participants of communicative acts can choose from different aspects of language. These include modality (the use of modal verbs and expressions), intonation and lexical elements that express the speaker's attitude, and are also dependent on contextual factors.

III. Factors that affect interpretation in technical discourse

3.1 Contextual limitations

Interpretation, or de-codification of meaning, is another important part of academic or other types of communicative acts. The interpretation of messages triggers past and present knowledge as well as other clues that we can recruit from the moment and situation where the expression is produced. Thus, the interpretation of "there is a crane in the park" may greatly vary depending on the situational and contextual elements i.e. it may be understood as an aquatic bird or as a mechanism for lifting heavy objects. Consequently, there are a number of factors that constrain the interpretation of an utterance (i.e. a linguistic expression in context) regardless of the communicative strategy used. Figure 2-1 shows these constraints.

In fact, when reading a text, we can use different resources that will help us activate what is informatively relevant for a specific "reading situation". Since trying to pay attention and interpret the full content continually would demand a high effort of the reader/listener. When interpreting a piece of technical discourse, we may draw on:

1. The co-text, or meaning of words or concepts surrounding one concept, often referred to as 'linguistic context'.
2. The previous discourse or immediate linguistic clues.
3. The clues from physical perception.
4. The socio-cultural context, or knowledge of social situations and interaction.
5. The 'cognitive' context, or background information stored in short-term and long-term memory.

Additionally, when interpreting a text, we often rely on:

6. The extended context, or previous knowledge of the field.
7. The meta-linguistic context, or knowledge of similar text types.

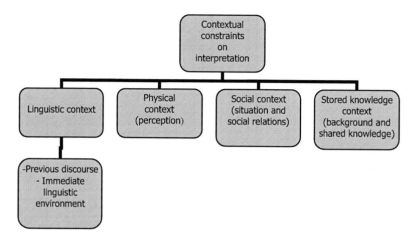

Figure 2-1 Constraints on interpretation

Summing up, context can influence communication in the following ways: by taking into account what was said before (previous discourse), the linguistic environment where the expression is produced and the type of discourse, i.e. the type of genre. Likewise, the type of register typology (formal or informal), the field of discourse (medical, economic, engineering, etc.), the environment that the participants can perceive through their senses (see, hear, smell) and consider as the physical context. Similarly, the social context concerning what is happening and the social relations between participants may introduce a further constraint. Finally, the previous knowledge of other or similar contexts that the participants have known or shared may influence the way the communicative act is understood.

3.2 Polysemy

A characteristic feature of scientific and technical discourse is the occurrence of polysemy. In *polysemy* a multiplicity of meanings may co-exist in the same term. The polysemic word contains a series of meanings that are semantically related to each other while also independent. Thus, the same word integrates different semantic layers. This is the main difference with *homonymy*, where two seemingly identical words in form and sound have no related meaning. For example, *die* is a noun designating an object-shaping device but also shares form with the verb "to die". However, these are distinct and separate words. In the case of polysemy basic schemas (*in/ out; up/down*, etc.) are shared and can be activated, as in *resistance* where the basic "opposition" schema remains active along its various senses.

As can be seen in the case of "bleeding" (figure 2-2), contextual clues (such as a picture and its accompanying text) may help to disambiguate this concept in ACE language.

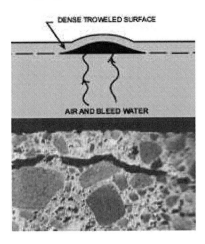

Figure 2-2 Contextual disambiguation of "bleeding"

> "*Blisters may form on the surface of fresh concrete when either bubbles of entrapped air or bleed water migrate through the concrete and become trapped under the surface, which has been sealed prematurely during the finishing operations. These defects are not easily repaired after concrete hardens*".

Table 2.2 shows another case of polysemy in "strain". This word has at least 3 different domains or "layers" where we can use it with their corresponding senses. What is more important, these senses have a connection. They are all linked by the basic meaning of 'reaction to a force'. Thus, by means of meaning extension, basic meaning gets extended to other domains, which also reinforces the principle of linguistic economy.

DOMAINS	ASSOCIATED SENSES
Architecture, construction and civil engineering • Structures • Mechanics	1. A change in length caused by a force. 2. The energy stored in an elastic body under load.
Medicine	To injure a part of the body by stretching it.
Economics/Politics	A difficulty caused in a system when its resources are severely tested.

Table 2-2. Polysemy in "strain"

3.3 Terminological clues

Ever since the development of English for Specific Purposes (ESP) in the 1970s, theorists in the field have analysed the terminological potential of specialized vocabulary. Herbert (1969:7) emphasized the importance of what he called 'semi-technical' words by saying:

"Much more difficult are the semi-scientific or semi-technical words, which have a whole range of meanings and are frequently used idiomatically; (...) words like plant and work and load and feed and force, words like these look harmless, but they can cause a lot of trouble to the student".

Other authors have also tried to draw the line between technical, semi-technical and general language vocabulary (Kennedy & Bolitho, 1984; Trimble, 1985), but, in many cases, this is a thin dividing line. Sometimes, the meaning of words from the general language is transferred to a technical domain (e.g. "mouse", the animal, or the computer accessory; "virus", in medicine and in computing). Also possible is the transference from a specialized field to a more general domain (e.g. His heart works like a broken "engine"), and even, there may be cases in which words are transferred and used across different specialized fields (e.g. "system", used in computing, and in telecommunications, or "pile", basically meaning `a heap of', and also 'a foundation type in building construction and in civil engineering).

As can be observed, the distinction between a "term" (specialized lexical unit) and a "word" (lexical unit commonly used in general language) is not easy to establish without proper analysis and observation. Because of this, most terminological studies nowadays aim to show and describe terms/words in their context of occurrence relying on statistical tools for the analysis. As we will see in chapter 4, these novel approaches to the study of terms are based on the approach called "corpus linguistics". By using computer programmes designed to that end, corpus linguistics analyses can describe the relations holding between a term/word and its immediate context.

3.4 Meaning through visual content

As seen in chapter 1, it is common in LACE to handle a large amount of visual input, in the way of diagrams, tables, graphs, pictures, etc., which can outnumber the linguistic information we receive, and help interpret meaning. Take the example in figure 2 above; aside from the rest of contextual clues that foster disambiguation in the concept of "bleeding", the accompanying image showing a layer of concrete in the process of setting intensifies its meaning, especially for the expert in the field. Furthermore, when asked to draw our conceptual image of a *house*, we may sketch a few lines that together provide a familiar representation. It would seem unnecessary to include many details in the picture, e.g. windows, door, etc... In other words, we would probably give a complete, and at the same time, basic representation of our mental configuration of "a house" to be easily interpreted. Visual content has a powerful impact on ACE audience, who commonly share a high degree of familiarity with visual representations. Likewise, it also serves to appeal to the immediacy attached to non-propositional, pre-verbal description. Let us remember the popular saying "a picture is worth a thousand words".

IV. Summing-up

In this chapter, we have seen that meaning does not take place out of context, but rather in specific situations that involve participants. We have also explained that meaning is grounded in physical experience, both in the production and in the interpretation stages. Meaning in LACE is influenced by various factors, such as terminological shifts or the contribution of visual information, which shape the ways in which content is produced and understood. Finally, in this unit we have outlined the role of context as an important element in the discourse of architecture and civil engineering.

V. References

CARTER, D. (1994). *Introducing Applied Linguistics: an A-Z Guide*. London: Penguin.
DURÁN, P. & ROLDÁN-RIEJOS, A. (2008). "The Role of Context in the Interpretation of Academic and Professional Communication". In Gibert, T & Alba, L. (eds.) *Estudios de Filología Inglesa*. Madrid: UNED.
HERBERT, A. J. (1969). *The Structure of Technical English*. Harlow: Longman.
KENNEDY, C. & R. BOLITHO (1984). *English for Specific Purposes*. London: MacMillan.
JOHNSON, M. (1987). *The Body in the Mind: The Bodily Basis of Meaning, Imagination, and Reason*. Chicago: University of Chicago Press.
LAKOFF, G. (1987). *Women, Fire, and Dangerous Things: What Categories Reveal About the Mind*. Chicago: University of Chicago Press.
LAKOFF G & NUÑEZ, R. (2000). *Where Mathematics Comes From: How the Embodied Mind Brings Mathematics into Being*. New York: Basic Books.
TRIMBLE, L. (1985). *English for Science and Technology. A Discourse Approach*. Cambridge: C.U.P.
WIDDOWSON, H. G. (1986). *Explorations in Applied Linguistics 2*. Oxford: O.U.P.

Follow-up practice unit 2

1. **Prepare 2 similar tables as in the "Strain" example given on page 21 with 2 appropriate architectural or civil engineering terms. For example, "stress", "resistance", "strength", etc.**

2. **Look for short fragments of ACE texts containing examples of polysemy (at least 5), mark them off, and indicate how these terms can mean something different in other contexts** (original source texts should be provided by inserting/ pasting them into this practice).

3. **Based on the previous task, identify the kind of terminological change these words/terms have experienced** (i.e. from general language to specific-field language, etc.).

4. **Present 3 examples of works of engineering and/or architecture that express some type of meaning because of their shapes or because of their interpretation.** For example, the shape of Calatrava's Alamillo bridge in Seville (Spain) may be interpreted as a harp or a fan, having to do with the river sound or to Seville's popular folk image. Include 1 photograph of each example.

Check it out!

To do this practice you may need to search and work with texts on line; you have different options:

1. Visit your library web page and select electronic journals, select topics of your choice and interest, and then download the files in full text PDF format to start working.

2. Subscribe free to Science Direct to have full access to these and other on line resources.

See: www.LACE.aq.upm.es for sample exercises from this unit.

CHAPTER THREE

COMMON GENRES AND TYPES OF TEXTS

In this chapter you will learn about:

✓ Specific details about common genres in ACE communication
✓ Their main conventions
✓ How and when to use them

Introduction

Genres do not only refer to different literary creation (e.g. novels, drama, poetry, etc.), musical types (jazz, folk, country music, ballads, etc.) or movies (comedy, suspense, horror, etc.). Genres are used to study communication, and represent conventionalized ways of communicating and of organizing discourse. As mentioned in Chapter 1, various genres can be identified in academic and professional language and we will see them in the next sections.

I. Genre. Understanding the concept

According to different authors, a genre is a standardized communicative event taking place in situations that could be defined as functional rather than social or personal (Evans, 1987). Genres could thus be explained as a sub-category of communicative events, i. e. text types, with some common communicative purpose or goal (Swales, 1990). Two concepts can be highlighted from genre definitions: "function" and "purpose". In fact, genres are characterized by being goal or purpose-driven and by being developed with specific function/s in mind. Genres are therefore ways to get things done using language in particular contexts. In fact genres, or discourse modes, are successful only when they employ conventions that other members of the scientific community know and share (Hyland, 2008).

Additionally, genres often follow three conditions:

- ✓ A rhetorical structure: each genre has an internal structure that makes it identifiable.
- ✓ Politeness conventions: each genre includes conventional courtesy communicative rules.
- ✓ Purpose of that specific genre: each genre is written for specific academic-professional purposes.

1.2 Considerations when writing up genre

On reaching university level, students face the convenience to write in a variety of "academic styles". The writing tasks students should engage on vary from one degree to another, but all of them are likely to partake in one basic feature, i.e. a certain degree of formality, which is common to most genres. Here are some factors to be considered when writing up academic genre.

- Audience: knowledge of prior expectations and experience in the field.
- Consider the purpose and strategy of the document.
- Organization: information presented to readers in a predictable format.
- Style/register: being consistent with the style required in any given type.
- Flow: keeping a smooth and clear connection between different ideas so as to help the reader follow the flow of the text. Using linking words and correct punctuation.
- Using correct paragraph segmentation:
  Paragraphs are units of meaning with a central or dominating idea and a number of ideas clustered around. Most of the times, as readers, we do not take the time to read through the whole texts, this is why well-written paragraphs allow the reader to skim-read for the gist, or scan for some pieces of information. Paragraphs serve several functions. Paragraphs may have different logical relations, among themselves and with the whole text, such as addition (*besides, additionally, in addition*), instantiation (*that is, for instance*), cause & effect (*consequently, thus, hence*), time (*first, second, then*) and space relations, comparison or conclusion (*in short, in sum, finally*).

- Presentation: double-check your piece of writing for mistakes in grammar, spelling or punctuation, and/or hand it over to a second person to get feedback. Consider the overall format.
- Title/heading/subheadings: is it representative/ is it neutral or biased?
- Visuals: relations between verbal & non-verbal language.
- View-point:
  - ✓ Is the text signed?
  - ✓ Does the writer take an overt/covert stance?
  - ✓ Is she/he detached/involved?
  - ✓ Is she/he a layperson/an expert narrator?

II. Common genres in ACE discourse

There are many different genres; the preference for some over others depends on our goals, needs and readership. In this unit we will focus on some that can be useful to university students in the field of architecture and civil engineering: formal e-mailing, summaries and abstracts, case-studies and the experimental research report.

2.1 Formal e-mailing

E-mails started in 1965 as a way for multiple time-sharing electronic communications. It was quickly extended to become *network e-mail*, allowing users to pass messages between different computers. It was Ray Tomlinson who initiated the use of the '@ sign' to identify the name of user as different from the rest of e-components in mailing addresses.

Generally speaking, e-mail users do not care too much about punctuation, grammar, spelling or any other potential errors. In principle, rules governing this genre are more flexible than in other genres. This is mainly due to the fact that the transmission of information prevails over any constraint of format or style. The bottom line idea is that e-mailing is meant to be ephemeral, functional and transitory. Nonetheless, when using e-mailing for academic or professional purposes, this type of document is to fulfill both legibility (format, length) and intelligibility (editing) conditions. As a general rule, although email correspondence may tend towards informality, it should follow the same principles as any other form of business correspondence.

Here are some basic conventions about style and content:

- In general email messages follow the style and conventions used in letters or faxes. You can use salutations such as: *Dear Mr...*, *Dear Jane*, and complimentary closes, such as: *Yours sincerely*, *Kind Regards*, *Best wishes*.
- Do not confuse personal messages with business messages. In a business message use the same rules as for a letter: write clearly, carefully, and courteously; you should consider audience, purpose, clarity, consistency, conciseness, and tone.
- Use correct grammar, spelling, capitalization, and punctuation, as you would in any other form of correspondence.
- Do not write words in capital letters in an email message. This can be seen as the equivalent of shouting and therefore has a negative effect. If you want to stress a word, put asterisks on each side of it, e.g. *urgent*.
- Keep your email message short and to the point.
- It always fares best to send 1 message for 1 topic. This helps to keep the message brief and makes it easier for the recipient to answer, file and retrieve it later.
- Check your email message for mistakes before you send it, just as you would check a letter or a fax message.
- In "second round" emails, it is acceptable to omit the salutation and the complimentary close when the sender and recipient have been exchanged previous messages.
- A good email sticks to the subject. Ideally, emails should be visible within one simple screen-view without scrolling down, and should follow the inverted pyramid content structure, with most important first and less relevant second. If in need of writing up more information, break your email into paragraphs which deal with one idea.
- Always mail back, even if you have nothing to respond at that particular time. Simply acknowledge the previous message and say that you will mail again (:"I just read your message").

Typical Structure:

Header including:
  Sender's name
  Date and time
  Addressee name
  E-mail topic or subject
  Enclosed attachments
  Body of the message
  Signature

> Header including:
>   Sender's name
>   Date and time
>   Addressee name
>   E-mail topic or subject
>   Enclosed attachments
>   Body of the message
>   Signature

Sample of formal email

To: Peter Fish
Subject: Refit of Lauren Road Store
Attachments: Plan of premises; Specification list; Architect's drawings

 With reference to our phone conversation this morning, I would like one of your representatives to visit our store at 43 Lauren Road, London NW3 4TN, to give an estimate for a complete refit. Please could you contact me to arrange an appointment?

As I mentioned on the phone, it is essential that work is completed before the end of March 1st, and this would be stated in the contract.
I attach the plans and specifications.

Alison Plant (Ms)
Assistant to K. Bellon, Managing Director
Superbuys Ltd,
Woolword Road, London SW 3 7DN
Phone.: 020 8444 6154
Fax: 020 8444 9543
alis.plantadvisory.com

2.2 Technical Summary

A summary is a short re-elaboration of a prior original text, it should keep its gist and at the same time condense its content.

Conventions to follow when writing a summary:

- ✓ Do not start by "the text is about..."
- ✓ Express its basic ideas with your own words.
- ✓ Provide a balanced coverage of the original.
- ✓ Present information in a neutral way. Be objective. Do not express your own views.
- ✓ Recreate the source text in your own words avoiding paraphrasing or copying.
- ✓ Do not improve the source by adding supplementary information.
- ✓ Go to the essentials. Leave out redundant, unnecessary details. Do not include quotes "...", specific data or figures unless highly relevant
- ✓ Write a draft: the summary result does not have to keep the original paragraph/information sequencing.
- ✓ The summary is independent of the main text.

> *A good technical summary is:*
>
> ✓ Brief : Maximum 20% of the original text
> ✓ Informative
> ✓ Understandable
> ✓ Independent of the text
> ✓ Complete: not in telegraphese
> ✓ No new material mentioned
> ✓ Adequate

2.3 Abstract

As explained in Chapter 1, an abstract (usual length about 250 words) is not equivalent to a summary, but rather it is like the DNA of a work of research (project or article). It is therefore a structured and self-contained piece of discourse. More importantly, frequently a scientific committee may decide out of an abstract if a proposal is accepted for a conference, a journal, a book or a project. If done well, it will make the reader want to learn more about your research, work, etc. An abstract often follows the internal content structure:

Background: (revision of previous relevant studies in the field)
Objective/s: (the goals of my study)
Methodology: (how I have conducted my research)
Results: (obtained from the analysis)
Discussion and conclusions

2.4 Case Study

In LACE, case studies can be effective methods to present representative cases of building structures of special interest. Case studies often include the following features:

✓ The use of representative authentic cases, therefore they should include some generalizable elements. For example, a case study about building the London Millennium Bridge is likely to mention the lateral vibration problems found during its construction, and the solutions applied, which may be useful for other engineers/architects in similar cases.

- ✓ They present in-progress authentic cases/materials already tested.
- ✓ They include precise facts and numerical data/quotations from experts.

Ex. 3-1 Millennium Bridge

And they present this rhetorical structure:

1. Presentation
2. Discussion of problems
3. Offer of possible solutions/Methodology
4. Results and lessons learned (Conclusions)

You can find examples of ACE case studies at:

http://www.springerlink.com/content/r6x457888jg5v4q3/

http://www.matscieng.sunysb.edu/disaster/

2.5 Experimental research report

The documents ascribed to this genre are written by an investigator to describe or advance a research study she/he has started up/completed. Some examples are:

- ✓ Controlled scientific experiments: empirical tests including as many factors as possible.
- ✓ Correlational studies including the comparison of variables.
- ✓ Questionnaire surveys.
- ✓ Computer-generated models.

Experimental research reports are characterized by:

- Being designed around a research question (sometimes implicit).
- Formulating a hypothesis in response to the central research question.
- Being quantitative: based on the study of numerical data/variables.
- Including tables, graphs, & diagrams.

As regards rhetorical moves, this genre follows this structure:
Title—Abstract (key-words)—Introduction Methodology—Results—Discussion—References

III. Summing up

In this chapter we have learnt about genre conventions and about some common genres used in the discourse of architecture and civil engineering. We have also learnt that these types of documents follow close-set guidelines and an internal—or rhetorical structure—with few variations. In particular, we have seen formal emailing, summaries and abstracts, case-studies and the experimental research report.

IV. References

DUDLEY-EVANS, T. (ed.). (1987). "Genre Analysis and ESP". *ELR Journal* 1.

DUDLEY-EVANS, T. (1988). "One-to-One Supervision of Students Writing MSC or PhD Theses", en A.E. Brooks y P. Grundy (eds). 1988.

—. (1989). "An Outline of the Value of Genre Analyis in LSP Work", LAUREN, C. *Special Language: From Human Thinking to Thinking Machines*. Londres: Multilingual matters.

DUDLEY-EVANS, T. & M.J. ST. JOHN (1998). *Developments in English for Specific Purposes. A Multi-disciplinary Approach.* Cambridge: C.U.P.

DUDLEY-EVANS, T. (2000). "Genre Analysis: A Key to a Theory of ESP?" in *Ibérica AELFE,* 2, pp. 3-11.

HYLAND, K. (2008) "Seminar on Genre" at Departamento de Lingüística Aplicada C. y T." Madrid: UPM.
SWALES, J.M. (1985). *Episodes in ESP*. Oxford: Pergamon Press.
—. (1990). *Genre Analysis. English in Academic and Research Settings*. Cambridge: C.U.P.
SWALES, J.M. & C. FEAK (2000). *English in Today´s Research World*. Michigan: University of Michigan Press.

Common Genres and Types of Texts

Follow- up practice unit 3

1. Complete the following tasks:

- Do you think e-mail is currently replacing other forms of communication? Write a short composition giving reasons for your answer (about 100 words)
- Write two samples of e-mail: (1) A formal one applying for a job, as a covering letter for your CV. (2) An informal one to a partner/colleague in another office from your company.

2. Read the following text and write a summary about it bearing in mind the characteristics of a good summary specified above:

AGGREGATES The material combined with cement and water to make concrete is called aggregate. Aggregate makes up 60 to 80 percent of concrete volume. It increases the strength of concrete, reduces the shrinking tendencies of the cement, and is used as economical filler. Aggregates are divided into fine (usually consisting of sand) and coarse categories. For most building concrete, the coarse aggregate consists of gravel or crushed stone up to 1 1/2 inches in size. However, in massive structures, such as dams, the coarse aggregate may include natural stones or rocks ranging up to 6 inches or more in size. The distribution of particle sizes in aggregate is determined by extracting a representative sample of the material, screening the sample through a series of sieves ranging in size from coarse to fine, and determining the percentage of the sample retained on each sieve. This procedure is called making a sieve analysis. For example, suppose the total sample weighs 1 pound. Place this on the No. 4 sieve, and shake the sieve until nothing more goes through. If what is left on the sieve weighs 0.05 pound, then 5 percent of the total sample is retained on the No. 4 sieve. Place what passes through on the No. 8 sieve and shake it. Suppose you find that what stays on this sieve weighs 0.1 pound. Since 0.1 pound is 10 percent of 1 pound, 10 percent of the total sample was retained on the No. 8 sieve. The cumulative retained weight is 0.15 pound. By dividing 0.15 by 1.0 pound, you will find that the total retained weight is 15 percent. The size of coarse aggregate is usually specified as a range between a minimum and a maximum size; for example, 2 inches to No. 4, 1 inch to No. 4.

3. Look for a prototypical example of the following genres: an experimental research report and a case study. Also provide: full reference and access details of the document, and state reasons why it adjusts to the model, or, if necessary, where it does not comply with the genre guidelines.

See: www.LACE.aq.upm.es for sample exercises from this unit.

CHAPTER FOUR

THE USE OF CORPORA APPLIED TO SPECIFIC LANGUAGE PURPOSES

In this chapter you will learn about:

✓ Corpus linguistics and its application to architecture and engineering texts
✓ How to compile your own linguistic corpus
✓ The use of AntConc software application

I. Introduction

In Latin, corpus (plural corpora) means "body". In current linguistics, a corpus is a compilation of texts in electronic-format representing a variety or state of language (Sinclair, 1991a; McEnery & Wilson, 1997). A corpus constitutes a standardized reference of the discourse that represents with regard to specific analytic purposes.

It should have a limited size, although the larger the sample, the more results we can obtain.

In contrast to previous approaches, which were based on the analysis of other lower-scale statistical variables, corpus linguistics aims at describing the main relations between a term/word and its immediate context, that is, corpus analysis shows how language functions, its use and how words relate to each other by giving out the "real" frequency of a term in its context. In fact, corpus linguistics is based on observation of huge amount of data, sometimes up to millions of linguistic items—here called "tokens"—that can be screened from a purely quantitative perspective. This "objective" view, coupled with the fact that populations can, in principle, be as large as research requires, has favoured and promoted the widespread adoption of corpus analysis in various scientific domains, and has made it necessary for any solid research to include results obtained from a linguistic corpus.

Through corpus linguistics we can examine and obtain both oral and written samples. For example, a possible study could be to study the main expressions used by engineers when presenting a report. Previously, we would have to decide the engineering domain we are going to study and its scope including the selection criteria for the *mining* of data. Therefore, with CL analysis, we are likely to bear in mind and to establish a previous objective. For example, we may want to check if the term *lining* appears in combination with *grout* or not. In summary, corpus linguistics allows to discuss data as well as to carry out quantitative and qualitative analysis.

As a basic condition to corpus analysis, the texts compiled must be arranged in electronic format, so that they can be tagged (labelled), classified, quantified and studied. Owing to this restriction, corpus linguistics necessarily depends on computer tools that allow carrying out all these functions. Wordsmith Tools and AntConc are two of them.

In this chapter, we will guide you through the steps to compile a corpus of texts touching on issues within your field of study or work, and we will also show you how to use AntConc, a free downloadable concordances programme that will allow you to handle large amounts of linguistic data. But first, let us start by searching and selecting the sources to build up the corpus.

II. Searching and filing the sources. Corpus compilation

As we have explained in the previous section, corpus analysis requires the linguistic data—that we will call samples—to be saved in electronic format. There are different ways to search and store these data. Electronic journals and materials are often accessible from your university computers (visit the library section in your School or Faculty or Subscribe free to Science Direct to have full access to these and other on line resources).

2.1 Instructions to compile the corpus

- Visit your School or Faculty Homepage, or similar website where you have access to electronic bibliographical resources.
- Enter the library section.

- Click on the electronic resources section and search for electronic journals (you may search by topic, by author or by title).
- Once the list of journals is displayed, select the ones dealing with topics within your fields of interest.
- Once the e-Journal is displayed, click on the title and follow the directions below:
    - Select 1 in the list: select 1 volume/ year within the field you plan to work on and click to open (1 Volume often includes 12 issues).
    - Click on the volume.
    - Click on 1 article (select the format pdf before opening).
    - The article opens in pdf format.
    - Using the right button on the mouse: "Select all" > "control + c", and save the file. Make sure the full text has been saved in .txt format.
    - Next, save each new file in .txt format, and also to give this file a specific name including, at least, the abbreviated name of the journal, the Volume number and the date, for instance: *Compostruct-08-07* (standing for *'Composite Structures', volume 08, year 2007*). It is often advised to compile texts dating back to the last five years.
    - We will follow the same procedure, using different volumes, issues and journals, until we have compiled a corpus of up to at least 30 articles. These files will be compiled in a folder tagged "corpus".

2.2 AntConc downloading and use

To carry out the second part of the practice, you will have to download the programme at http:// www. antlab.sci.waseda.ac.jp/software.html (tutorials are also available on this web site). The programme is downloaded free and creates a folder directly on your computer desktop.

Once the corpus has been compiled, open AntConc and follow these steps:

- Double click on the program.
- Upper left hand corner: click on "file".
- Open file or Open directory, in this case "corpus".
- Once all the files have been opened (left hand side), AntConc includes the following functions:

- Activating the "concordance" function, you find all the examples of a term and its frequency (number of times it appears in the full corpus). In case you are looking for a specific term, enter the word in "*search term*" (lower left section) and then click on "*start*" (See Ex. 4-1 below for a sample of the concordance lines of *structure*).

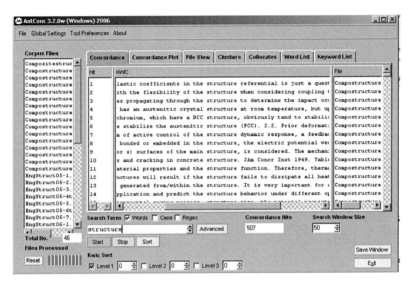

Ex. 4-1 Concordance lines for *structure*

- Clicking on a term in blue as it appears in the "concordance" lines, the "view file" function displays the term in context.

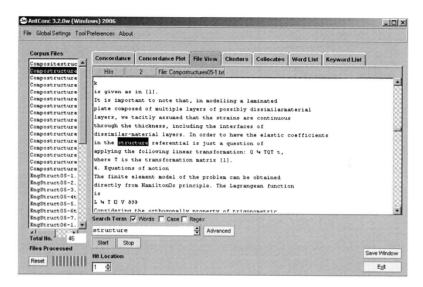

Ex. 4-2 Sample context for *structure*

- To know the total volume of the corpus (total number of word tokens) and the total number of different words, activate "*wordlist*". The frequency and distribution of all the words in the corpus will be displayed. In Ex. 4-3 below, the total number of word tokens amounts to 262,086, and the total number of different word types (different word categories) is 13,271.

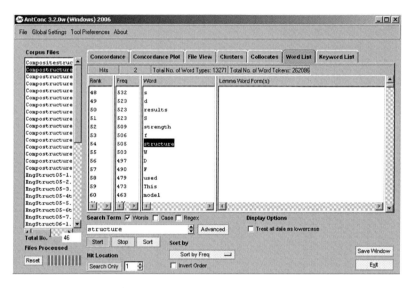

Ex. 4-3 Total number of word tokens and different word types

The programme also shows the total number of files compiled (left hand section), in this example we have 46 different files corresponding with 46 different full-text scientific journal papers

- Finally, activating the *"clusters"* function, the programme displays the different combinations of one word with other words in the total corpus. This function is especially useful in some the technical fields where compounds are common practice. See clusters for *structure* in Ex. 4-4. It is important to note that, in this case, some of the combinations are not relevant, since they include the definite and indefinite articles (examples number 1 and 3) or the verb *to be* (example number 2), these are not compounds but common word combinations. However, examples number 4 and 6 reveal clusters which may probably be highly representative in the corpus (namely, *supporting structure* and *primary structure* respectively), and thus, in the field we are analysing. To test this finding, further search would be necessary.

The Use of Corpora Applied to Specific Language Purposes 43

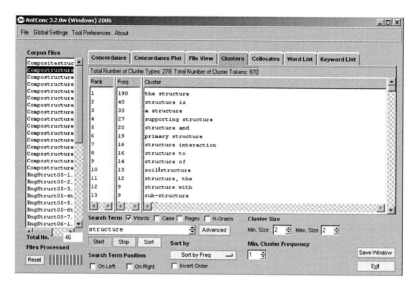

Ex. 4-4 Sample of clusters for *structure*.

III. Applications of Corpus analysis in the learning of specific vocabulary

Corpus analysis allows us to:

- Generate a corpus with terms that are field-specific and representative in our speciality.
- Provide statistical (quantifiable) data about the frequency and distribution of terms in the corpus.
- Handle and work with huge amounts of linguistic data in a simple accessible way.
- Offer the possibility of saving all concordance lines (the actual examples) of a term in a separate file for later consultation (select text and in "file" choose "save output to text file").
- Show the contextual realization of a term, that is, it describes a term in its context, and thus...
- This type of analysis promotes the learning of new vocabulary through the use of contextual information.
- Get familiar with frequent collocations and keywords used by our community discourse

IV. Summing up

In this chapter we have explained that linguistic analysis in recent years is mostly based on corpus results when it comes to disseminating new findings or to setting up empirical research. We have also explained the procedure to compile a corpus with articles from journals dealing with specialized topics. Finally, we have shown the functions of AntConc and the ways in which this programme can be used to extract specialized lexical units, or terms, to get a more realistic perspective on language for specific purposes. In short, the practice in this chapter is meant to be the first step towards the creation of a larger—ongoing—corpus that will help you increase the knowledge and the practice of specialized vocabulary. Likewise it will allow you to handle ACE communication better and more practically.

V. References

BIBER *et al.* (1998). *Corpus Linguistics. Investigating Language Structure and Use.* Cambridge: CUP.
BOWKER, L. & J. PEARSON (2002). *Working with Specialized Language. A Practical Guide to Using Corpora.* Routledge: London & New York
HUNSTON, S. (2002). *Corpora in Applied Linguistics.* Cambridge: CUP.
KENNEDY, G. (1998). *An Introduction to Corpus Linguistics.* Londres y Nueva York: Longman.
MCENERY. T. & A. WILSON (1997). Corpus Linguistics. Edinburgh: Edinburgh University Press.
SINCLAIR, J. (1991). *Corpus, Concordance, Collocation.* Oxford: OUP.

Follow- up practice unit 4

In this chapter, you will have to compile your own corpus with a least 30 articles extracted from journals within your field and complete the following steps:

I. Before using the concordances programme, write a list with the terms you anticipate as important in your field or specialty?
II. Indicate whether they correspond to the results obtained using Antconc. Write a list including the terms that match both cases.
III. Draw up a table including the 25 most frequent terms obtained with this tool, indicating as well their rank in the general corpus and their rank in your final corpus results. Remember that in your final list you will have to eliminate functional items like prepositions and conjunctions, and also "empty words" with general meaning that are not terms properly. Remember as well that to obtain better results from the search, you may activate the "clusters" function in the programme to get compounds or terms including more than one word, i.e, *thermal loss, energy-saving, reinforced concrete*, etc., which are highly informative and widespread in most technical fields.
IV. Assess this experience in terms how it can help to consolidate your knowledge of specialized vocabulary.

Shortlist of e-journals that could be used in this practice:

1. Applied composite structures
2. Applied thermal engineering
3. Automation in construction
4. Building and environment
5. Composite structures
6. Construction and building materials
7. Costal engineering
8. Engineering structures
9. Fire safety journal
10. International journal of fatigue
11. Journal of constructional steel research
12. Materials and design
13. Materials today
14. Structural safety

See: www.LACE.aq.upm.es for sample exercises from this unit.

CHAPTER FIVE

TEXTUAL STRATEGIES: LANGUAGE AND USE

In this unit you will learn about:

✓ Communicative strategies used in ACE texts
✓ Types of strategies: thematic progression, adverbs and modal verbs use, sentence arrangement, hedging.
✓ The strategic use of the passive voice in academic and professional texts

Introduction

As we have discussed in previous chapters (see 1 & 2), ACE discourse can be used for various purposes, e.g. to inform, to describe, to give instructions, to persuade, etc. To achieve these goals, authors can make use of linguistic strategies. These strategies may be either stylistic—associated with style and register—or linked to meaning. Both types appear interconnected in discourse. In this chapter, we will focus on some of the most frequent and relevant strategies used in ACE written communication.

I. Communicative strategies in LACE

1.1 Thematic progression

Let us consider some communicative strategies that clearly reflect the author's intention by highlighting specific discourse elements. The organisation of message, for example, typically follows the structure: "known information comes first while new information comes last". This development is often defined as thematic progression and its parts are known as (a) theme: beginning and (b) theme: rest of the sentence/message. This recurrent thematic progression, or way to organise a message, is a result from both cognitive and linguistic restrictions. On the one hand, from a cognitive point of view, speakers need some time to activate "new information", and thus by placing known information first, they get extra time to de-construct the message, progressing from the "known to the new". On the other hand, we tend to place syntactic heavy parts of the message at the end of the sentence to make its main idea more recent and easier to follow. This principle is known as "end-weight", that is to say, weightier parts in meaning and syntax are placed at the end of the message. Look at example (b) where the second part of the message from *required* onwards carries more weight than the preceding part. In addition it sounds more natural than example (a):

(a) That all new students enrol in the course register within the first two weeks is required.

(b) It is *required* that all new students enrol in the course register within the first two weeks (end-weight).

Furthermore, the theme function tends to contain known or "given" information and typically goes at the beginning of the sentence. It provides the topic of the sentence. Given information presents the following characteristics:

- It has been mentioned before.
- It can be easily inferred from the context.
- It can be easily understood from previous background knowledge.

By contrast, the rheme usually communicates new information, and therefore receives more focus of attention, special speech intonation, and is placed later in the sentence. Furthermore, rhemes cannot be usually deduced from the preceding discourse.

Look at the following example from a text explaining the advantages of pre-cast concrete, the given (theme) is in blue and the new information (rheme) in red:

The precast prestressed structural and nonstructural concrete units are usually in production prior to and during site and foundation preparation work. This enhances the speed of construction as upper-floor structural elements can be in production while site and foundation works are in progress.

As we can see, *This* refers to the process of preparing precast units. In this fragment, new information was contained in the preceding rheme, so it can be easily inferred or retrieved. The question of assigning focus (the most informative element in the message) is going to be mentioned again through the active-passive voice alternation. In addition, there are other typical features of scientific English, like the use of modality for the expression of attitude often realized in language by adverbs in sentence initial position, i.e, "obviously, frankly, arguably, apparently", etc, or by modal verbs, such as "may, might, should, would", etc. These are communicative strategies at the disposal of the scientific community, and are frequently used, either conscious or unconsciously.

1.2 Hedging

A particular discourse strategy commonly applied in academic and professional discourse, and therefore in ACE language, is hedging. According to Varttala (2001), hedges are "linguistic devices indicating less than complete certainty or commitment regarding information put forth". In other words, they are able to modulate part of the meaning expressed in the text. In everyday communication, hedges are used to express different functions, especially in politeness and acts of requesting, ordering, inviting, declining requests, etc. For example, hedging in academic discourse is frequently used as a linguistic strategy by the author of papers, reports, etc. (see examples below). Holmes (1984) points out that the main function of hedges is "attenuating negatively affective speech acts" and talks about "downtoners". In this sense, hedges are also known as "mitigators", because they are ways of protecting assertions and of being cautious in academic writing, e.g. avoiding the consequences of a negative statement.

Roughly speaking, two main types of hedges can be identified (Hyland 1998):

a) Content-oriented, i.e. about the precision of the text, and it is writer oriented (this includes academic courtesy and may serve for protecting claims). For example, the author may use the plural pronoun "we", instead of "I" not to appear arrogant by excessive personalising. The use of modifiers, such as "approximately, roughly, fairly", etc., may also attenuate the expressed content.

b) Reader-oriented, i.e. to create rapport or complicity with the reader and to avoid open criticism. For example, an article may say: "earlier studies have not tackled this question" instead of providing the actual authors' names.

1.2.1 Lexical hedges

Lexical items used for hedging can be divided into two broad categories:

(a) 'Verbal'
(b) 'Non-verbal'

'Verbal hedging' includes modal verbs (*may, might, can, could, would*), other lexical verbs such as 'appear', or 'suggest'; tentative cognition verbs (*believe, think, consider*), tentative linking verbs (*seem, become*).

'Non-verbal hedging' can be sub-divided into nouns (*suggestion, claim*), prepositional groups (*to our knowledge, in our opinion*), probability adverbs (*probably, possibly seemingly*), adverbs of indefinite frequency (*usually, generally*), adverbs of indefinite degree (*roughly, approximately*), probability adjectives (*likely, possible*), approximative adjectives (*about*).

Appropriate use of hedges in academic discourse requires a considerable degree of practice. Hedges primary function should be politeness and academic courtesy, e.g. to show respect for other scholars' work, rather than to avoid commitment or to escape responsibility in research.

1.3 Passivization

Passive voice use is very frequent in technical literature. In ACE language, passive voice usage seems a widespread convention regardless of the presence or absence of an agent (i.e. full or truncated passive). Consider the following examples:

(a) *Another tunnel has been built by Ferrovial tunnel division* (full passive: agent present)

(b) *Five new T-beams have been cast at the site* (truncated passive: no agent)

Truncated passives are more frequent than full passives in ACE, since the procedure (the action) is likely to be considered more relevant that the agent. Additionally, very often the agent does not need to be mentioned, it is irrelevant in the sentence. Nevertheless, truncated passives could be easily replaced by their active counterparts with subjects we/I. For instance:

(ci) *The actual material savings can be analyzed by considering two basic structural elements: a typical floor slab and a typical floor beam sample.*

(cii) *We analyzed the actual material savings by considering two basic structural elements: a typical floor slab and a typical floor beam sample.*

Since in ACE language the agentive subject is not usually human, the general use of the passive voice seems justified. Look at example marked (d):

(d) *The finished structure will be subject to less shrinkage forces and consequential cracking damage in the building frame.*

Converting the sentence into active and using *forces* as the subject would not sound as natural as in the passive example and moreover the focus of the sentence would be affected. Also, as seen above, the tendency in English is to place the given part at the beginning of the sentence (theme/topic) and the new part (rheme) later. Therefore, the passive voice looks as an important strategy in discourse since it enables us to place the agentive subject at the end of the sentence. In this way, the agentive subject would fall on a focal position. Look at the following example:

(e) *In prestressed concrete, predetermined forces are imposed on the concrete structural member by special high-tensile steels prior to the members functioning as a load-supporting element.*

The text marked in red highlights the logical subject (agent), placed in rheme position, and therefore receiving more focus, whereas the theme (affected subject) in blue is placed as given information. Look at a similar example:

(f) *Significant savings in materials can also be realized by precasting and prestressing techniques.*

However, very frequently, as we have seen in truncated passive, the agent is omitted. For instance, see below being the theme in blue and the rheme in red:

(g) *Both conventionally reinforced and prestressed concrete elements can be fabricated as precast units.*

In short, the general characteristics attributed to the passive voice are:

- ✓ It can be converted into a transitive active construction.
- ✓ They present an affected subject, but not always the agent (truncated passive).
- ✓ They consist of finite clauses.
- ✓ The verbs involved are usually dynamic; therefore there is activity and punctuality involved.

In addition, we can observe that the type of genre seems to have an influence on the choice of the passive voice. In abstracts, for example, passive voice use seems to respond to polite academic reasons, perhaps because of the conventions of the abstract genre (see chapters 1 & 3). A report, however, usually makes reference to some sort of action. The choice of passive constructions within this genre is more common than in the former case.

Everything considered, passivization is one of the syntactic strategies that can be used to assign focus. The two major positions in the clause are the beginning and the end. In the active sentence, the starting point for the message, (often called "the theme"), matches the "Agent as Subject" construction, while the "Affected participant" is in final position and receives end-focus.

Example (1) *The T-beams supported the roof.*

Subject: theme   Affected: end-focus

In the passive construction this arrangement is reversed. The Affected now provides the point of departure, coinciding with Subject, while the Agent takes the final position and receives end-focus.

Example (2) *The roof was supported by the T-beams.*

Affected subject: Theme   Agent: end-focus

Therefore, the active-passive alternation allows the speaker to arrange the message so that new information may be placed in end-position, while the element considered to be given or known may be placed in initial position or the other way round.

In conclusion, the extended use of the passive voice in technical discourse responds to the following reasons:

1. Absence or lack of interest in expressing the agent.
2. Thematisation or topicalization: end weight; end focus.

II. Summing up

In this chapter, we have seen that there are strategies a writer can adopt to make technical communication more effective. Among them are the preference for certain content organizations, such as a thematic progression in which known information comes first while new information comes last in the utterance, the frequency of the passive voice and the use of hedging devices. These attenuate the potential negative effects of a statement that could otherwise be "threatening" to the interlocutor or to the reader. We have also seen that strategies use in discourse responds to cognitive and strategic reasons that are reflected in grammar.

III. References

HOLMES R. (1997) "Genre analysis and the social sciences: An investigation of the structure of research article discussion section in three disciplines. *English for Specific Purposes*, 16: 4: 321-337.

HYLAND, K (1998) *Hedging in Scientific Research Articles*. Amsterdam, The Netherlands: John Benjamins.

RODEAN, L. (1981) "The Passive in Technical and Scientific Writing". In: Lally, T. (ed.) *JAC Volume 2, Issue 1/2* .Illinois State University.

VARTTALA, T. (2001) *Hedging in scientifically oriented discourse: exploring variation according to discipline and intended audience.* Available at: http://acta.uta.fi/pdf/951-44-5195-3

Follow- up practice unit 5

1. Provide a selection of hedges used in 2 ACE journal articles of your choice. Point out the most frequent ones, and explain their main function. Remember to include here the original contexts/fragments in which they appeared and to include the source/s (See chapter 4, on Journal articles downloading).

2. Look at the two examples of genres: abstract and report, and analyse the communicative strategies used in their writing. Write a comment on: (a) hedging and (b) passivisation (250 words). Provide reasons for your analysis.

Text 1: Abstract

(1) Elk Creek Dam in southwest Oregon is the only Corps of Engineers dam to be constructed using RCC that has passed beyond the planning stage to engineering and design. (2) Design and construction concepts for Elk Creek Dam are considered to constitute the current Corps of Engineers criteria for RCC for dams. (3) Because of the size and purpose of the structure, design concepts were established to assure a low rate of seepage through the dam.

Text 2: Report

(1) Due to the potential for significant downstream damage in case of overtopping, the dam was designed to control the probable maximum flood (PMF). (2) The control was provided primarily by storage of a high percentage of the PMF volume. The resulting design called for a dam that would have been 200 feet high (61 m). (3) In the spring of 1982, construction was begun at the site, with excavation, foundation treatment activities, and fabrication of pipe, gates, and other appurtenances. (4) In May 1982, construction of Middle Fork Dam was halted as part of a decision by Exxon to redirect development efforts at the Colony Project.

See: www.LACE.aq.upm.es for sample exercises from this unit.

CHAPTER SIX

METAPHOR IN THE DOMAINS OF ARCHITECTURE AND CIVIL ENGINEERING

In this chapter you will be learn about:

✓ Metaphor as an extended phenomenon in thought and language
✓ The role of metaphor (conceptual and linguistic) in ACE communication
✓ The use of metonymy in ACE contexts

Introduction

Metaphor has traditionally been associated with the world of literature and poetry. However, metaphor's function goes beyond lyrical creation. Recent research has shown that metaphor is a powerful cognitive tool working both in thought and in language whereby facilitating the transmission and comprehension of complex concepts and ideas. As linguists put it: "Metaphor takes a range of various forms (...) it has a surprisingly large number of functions, cognitive, social, affective, rhetorical and interaction-management" (Cameron and Low, 1999: 91).

Metaphor works as a conceptual mechanism enabling to establish similarities, i.e. we attribute to a concept that belongs to a certain conceptual domain qualities or elements that are characteristic of another conceptual domain. Correspondences between the attributes of domains are then established.

We speak about a source domain (usually more concrete and more direct to experience) and a target domain (more abstract), We also speak of mappings, or transferences, across them (Lakoff and Johnson, 1999). For instance, when referring in LACE to *concrete bleeds*, we are using a mapping that associates an inert material "concrete" (the target, more abstract) with a living entity (the source, bodily related) with this substance (blood). We do it in this way for two reasons, first because we are specifying a problem that concrete may have and that the engineer has to solve, and second to designate that concrete may be losing more water than desired. We know that concrete has no blood as a substance that circulates in living creatures, but accept the convention. If engineers speak in these terms it is because the conceptual metaphor +CONCRETE IS A LIVING BEING+ is recognizable and frequently used within the engineering community.

II. Role and use of metaphor in ACE contexts

Metaphor is a familiar phenomenon both in thought and language. It is frequent in everyday conversation and also in special types of discourse: scientific, technical, religious, etc. From the point of view of cognitive linguistics, far from being a complicated process, metaphor assists mental operations. Cognitively speaking, metaphor is a "mechanism of analogy where a concept belonging to a particular conceptual domain is conceived according to another one" (Lakoff and Johnson, 1980). Correspondences and mappings are established between attributes. In this sense, a source domain with mapping attributes and a target domain may be identified, as well as correspondences between them, as shown in the following Ex. 6-1.

Metaphor in the Domains of Architecture and Civil Engineering 59

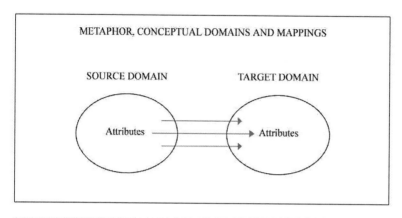

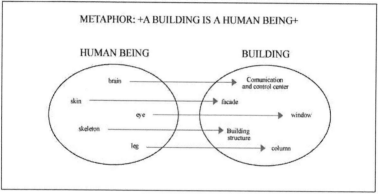

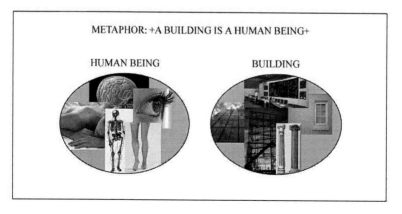

Ex. 6-1 Metaphor Mapping

This process primarily takes place at the level of thought, and is based on schemata and imagery evoked by human perceptual and personal experience (Johnson, 1987). The use of such process is widely extended and unconscious, permeating language and thinking. In Meta*phors We Live By,* Lakoff & Johnson (1980: 3-21) reveal some common conceptual metaphors that we use in everyday language:

ARGUMENT IS WAR[2] (i.e. to argue is to fight): Your claims are indefensible.
TIME IS MONEY: You are wasting my time.
HEALTH AND LIFE ARE UP: He's at the peak of his health.
SICKNESS AND DEATH ARE DOWN: He came down with flu.
CONTROL IS UP: I am on top of the situation.
BEING SUBJECT TO CONTROL IS DOWN: He is under my control.
MORE IS UP: The number of books printed every year keeps going up.
LESS IS DOWN: If you are too hot, turn the heat down.
THEORIES ARE BUILDINGS:

    a. Is that the foundation for your theory?
    b. The theory needs more support.
    c. We need to construct a strong argument for that.

In the experience of architects, it is not unusual that a building and a town are conceived as human beings, so corresponding conceptual metaphors would be: +A BUILDING IS A HUMAN BEING+[1] or +A BUILDING IS A BODY+, and +A TOWN IS A HUMAN BEING+. Here are further examples of conceptual metaphor in ACE contexts:

A BUILDING IS A LIVING ENTITY (Roldán-Riejos, 1995)
STRUCTURES HAVE RELATIONSHIPS (Roldán-Riejos, 1999)
EXISTIR ES MANTENERSE ERGUIDO (EXISTING IS STANDING UP) (Boquera, 2005)
LAS ESTRUCTURAS SON SERES HUMANOS (STRUCTURES ARE HUMAN BEINGS) (Santiago López, 2007).

---

[2] Metaphoric structures are conventionally written in capital letters.

2.1 Conceptual Metaphor

Metaphor may be used to express basic ideas in technical disciplines. In ACE communication, we find many metaphors expressing the malfunction of engineering materials or structures. Starting from the mapping +ENGINEERING MATERIALS/STRUCTURES ARE LIVING ORGANISMS+ both materials and structures are conceptualised as being affected by diseases and medical problems that the engineer ought to treat. This appears to be a recurrent metaphor in civil engineering. Therefore, it is easy to find examples such as: *concrete bleeds* and *concrete fractures*, meaning an excess of water and when concrete cracks; in addition *structures undergo fatigue, strain, stress*, they can also *age* and have a limited lifespan. These expressions and similar ones are borrowed from the source domain of biology and medicine and are extremely frequent in technical discourse, since they can appear in practically any genre: report, instructions, procedures, etc. They are conventionalized within the engineering community, although out of this community they may sound rare.

Conceptual metaphors can constitute complete semantic fields, unlike image metaphors, which simply project the schema of a mental image over another one. Throughout history, architects have displayed their creativity, both in individual works, such as buildings, and collective examples, such as the transformation of cities. Their works are a reflection of the author's personal perception and social environment in which they operate. Consider the following present-day examples:

The first one is a modern edifice built as a dental clinic and residence in Japan. This clinic is conceived as a conceptual metaphor. The image metaphor would be: +THE CLINIC IS A TEETH+. The architect provides a mirror image of a building design with the clinic function. [Building: Dental clinic Date: 2011 Architect: Hiroki Tanabe City: Nagano Japan/Tipology: Dental clinic].

Ex. 6-2 Visual metaphor provided by the architect: Teeth

Another example is represented by the International Neuroscience Institute, a unique facility whose shape is designed by the architect to resemble the human brain and the interior in its web page http://www.ini-hannover.de/en/home.html is said "along the horizontal section is a design similar to the structure of a human brain". [Building : International Neuroscience Institute INI /Date of construction: 2003 Architect: Cegid head office, Lyon GIB 8.4City: Hannover (Germany) Typology: Medical Center].

Ex. 6-3. Visual metaphor provided by the architect: Brain

Both buildings reproduce visual metaphors linking shape and function. In both examples the architects overlap perception and shape.

2.2 Linguistic Metaphor

In the business world, we can speak of "rocketing prices" because we see the cost of living as moving fast and rising very high, similar to the way rockets go up into the atmosphere. Thus the underlying metaphor would be: +PRICES ARE ROCKETS+. Equally, we can find the following expressions: "oil prices soar" or "stock shares plunged". Through the direct experience that we all share - things going up or going down- (source domain), we refer to the more abstract notion of price fluctuation.

Linguistic metaphoric expressions like these may be identified attending to an underlying conceptual metaphor, given that we have characterized metaphor as a cognitive tool being present both in thought and language. In addition, metaphorical expressions can be studied at a linguistic level, i.e. mostly focusing on their textual function, according to genre and to their function in discourse.

Discourse studies of metaphor usually catalogue linguistic metaphor as a type of figurative speech (Littlemore and Low, 2006, Deignan, 2005) in particular attending to its use. Cameron (2003) provides a clear definition of linguistic metaphor: "the use of a word or phrase that brings (or could bring) some other meaning to the contextual meaning". According to this definition, the word or phrase that brings "the other meaning" that contrasts with the main topic of the text is the Vehicle and the main theme of the text would be the Topic. For practical reasons, a more basic and concrete meaning is used to designate the vehicle. For example in: "Beam fatigue testing system", a *system to test beams* would be the topic and the *possible beam disease* would be the vehicle.

To identify linguistic metaphor in a text, two ways have been proposed. One is by identifying the vehicle (Cameron 2003, CREET group metaphor analysis) and requires two conditions:

(i) The meaning of a vehicle word or phrase should contrast with the topic domain, in the preceding example, *the mechanical system to test* contrasts with a *possible disease of the beam*.
(ii) A link or connection of meaning should exist between the contextual topic and the vehicle. In the example above the link between them would be *the need to test the physical condition of the beam*.

The second one is by identifying individual words that may have metaphorical meaning. In this option, the following steps are recommended:

1. Read the entire text/discourse to establish a general understanding of the meaning.

2. Determine the lexical units in the text/discourse.

3. (a) For each lexical unit in the text, establish its meaning in context, i.e. how it applies to an entity, relation or attribute in the situation evoked by the text (contextual meaning). Take into account what comes before and after the lexical unit.

    (b) For each lexical unit, determine if it has a more basic contemporary meaning in other contexts than the one in the given context. For our purposes, basic meanings tend to be:
    more concrete; what they evoke is easier to imagine, see, hear, feel, smell, and taste
    · related to bodily action
    · more precise (as opposed to vague)
    · historically older
    Basic meanings are not necessarily the most frequent meanings of the lexical unit.

    (c) If the lexical unit has a more basic current/contemporary meaning in other contexts than the given context, decide whether the contextual meaning contrasts with the basic meaning but can be understood in comparison with it.

4. If yes, mark the lexical unit or group as metaphorical

Table 6-1. Steps used by Steen (1999: 57) and the Pragglejaz group (2007) to identify metaphors.

2.3 Metonymy

Another important variety of figurative language is *metonymy*. Metonymy is also a cognitive tool that we use in thought and language. The difference with metaphor is that it does not *transfer* meaning in two different domains giving a relation of similarity, instead it *combines* meanings in the same domain. Therefore in metonymy an association between two entities of one standing for the other in the same domain is established. This relation can be a PART/WHOLE one, for example when we say "The current brain drain in Spain is due to unemployment", we refer to people with skills and knowledge that leave their country to work abroad, even if we just focus on one part of the body, the brain.

It can be said that metonymy highlights contiguity between two concepts, whereas metaphor works by establishing a similarity between them. As mentioned above, in metonymy there is only a conceptual domain not two, so the mappings are established in the same domain (see figure 2) where PR is considered as the point of reference and ZA is the Active Zone.

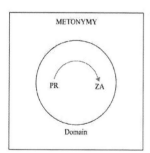
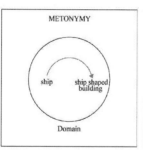
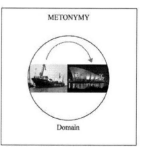

Ex. 6-4 Metonymy Mapping

Metonymy basically consists of conceptual associations that rely on world knowledge and connotations. Therefore, in a given conceptual mapping there is one conceptual entity (the vehicle) that evokes another conceptual entity (the target). Contiguity relations can be linked in causation, for example, "macadam" is a type of road pavement that takes its name from John Mc Adam, the engineer that invented it. In addition, a special case of metonymy is known as synecdoche, by which there is an inclusion of elements, for example in "Madrid has agreed on signing the treaty", "Madrid" is the vehicle that stands for the Spanish Government (the target), which is located in Madrid, showing a part/whole or whole/part relation. Thus, metonymy works by calling up a domain of usage and an array of associations.

Other examples of common metonymies are (Lakoff and Johnson, 1990: 38-39):

PRODUCT FOR PRODUCER
    He bought *a Ford*.
    He's got *a Picasso*.

OBJECT USED FOR USER
    The *sax* has the flu today.
    The *buses* are on strike.

CONTROLLER FOR CONTROLLED
    Napoleon lost at Waterloo.
    A Mercedes rear-ended *me*.

INSTITUTION FOR PEOPLE RESPONSIBLE
    Exxon has raised its prices again.
    You'll never get the *university* to agree to that.

THE PLACE FOR THE INSTITUTION
    The *White House* isn't saying anything.
    Wall Street is in a panic.

THE PLACE FOR THE EVENT
    Remember *the Alamo*.
    Watergate changed our politics.

In Architecture and Engineering, there are many examples of the metonymic relation CAUSE-EFFECT. The architect or the engineer work transforming the environment: they create bridges, dams, buildings, ports or highway systems and they integrate them in a specific area. According to this, the metonymy they use would be visual and included in the end-result of their work. For example, a building by the sea to lodge a hotel, flats and a shopping centre with the shape of a shell or a bridge that will open a neglected part of the city with the shape of a doorway, etc.

Sometimes, metaphor and metonymy can be at work in the same figure of speech, in other words, we could interpret a phrase metaphorically or metonymically. Even though metaphor and metonymy are quite different in the way they map ideas, they can work together in a continuum where it is difficult to say which is which. This phenomenon has been named as "metaphonymy" (Goossens, 1995). For example, in "The London eye opens daily from 9 am to 9 pm" there is the metaphor established by the substitution of the wheel by an eye (similarity relation). And there is a metonymy by taking the whole (the name of the attraction) for the part (people working there) in a contiguity relation.

Another example in the field of architecture showing the coexistence of these two linguistic phenomena would be in the expression: "Moneo's cubes have raised controversy". It was a headline taken from a newspaper when The Kursal Building was beat in San Sebastian. The interpretation being: the cubes represent the building + THE FORM FOR THE WHOLE +, and secondly, there are two conceptual metaphors + THE BUILDING IS A HUMAN BEING + (personification) and CONTROVERSY + it has been considered as an object capable of being lifted, at the same time providing an orientational metaphor where UP can be considered MORE. [Building : Kursaal /Date of construction: 2003 Rafael Moneo City: San Sebastian (Spain) Typology: Pavilion of exhibitions].

Ex. 6-5 Visual metaphor provided by the architect: Cubes

2.4 Visual metaphor

One of the important skills that both architects and engineers should acquire during their training is verbalising, that is, putting into spoken and written words their mental images and representations, which may explain why metaphor or metonymy are common features. Caballero (2003) has examined the occurrence of conceptual metaphor and, particularly of image metaphor in architecture. Many of the examples that she quotes from her corpus of texts have a visual origin and nature. Equally, she points out that it is not often easy to differentiate conceptual from image metaphor in architecture texts, claiming that we can observe an interplay between both because of "the visual and aesthetic constraints of the discipline" (2003:150). Additionally, we also underline the key role of the visual component in engineering, as well as the aesthetic element, which is frequently underestimated. In fact, this aspect can be proved in many historic examples, such as Segovia aqueduct or the Chinese wall. In Roldán and Úbeda (2006: 538) additional evidence shows a considerable number of metonymic images used in descriptive engineering texts.

Visual metaphor may be influenced by context in the same way as conceptual metaphor is influenced by topic (Kövecses, 2009). Thus, in the sentence: *The architect Renzo Piano docks with a new public building in the city of Amsterdam*, the writer chooses *docks* as a link with the ship-shaped building that is located at the port of the city of Amsterdam. Also the architect Renzo Piano is seen as ship that docks in a harbor with his new building and this building design takes the shape of a ship: +THE ARCHITECT FOR THE THING+ metaphor. [Building : NEMO (New Metropolis)Date of construction: 1997 Architect: Renzo Piano City: Amsterdam (Holland) Typology: Museum].

Metaphor in the Domains of Architecture and Civil Engineering 69

Ex. 6-6 Visual metaphor provided by the architect: Ship

III. Summing up

In this chapter, we have approached metaphor from a cognitive perspective. Metaphor is considered a general phenomenon with cognitive underpinnings that involve language and thought. In the technical fields, very frequently concepts are transferred across different experiential domains so that abstract ideas are brought to a more down-to-the-ground level. Engineers and architects think of problems that affect structures as patients to be treated and they look at themselves as doctors that treat them. This includes examples like: "Pathology of structures"; "auscultation of dams"; "aging in buildings", etc. Language and communication among architects and engineers is seen as a very powerful tool, where conceptual, image, metaphor or metonymy play a key role in work activities; this is a new dimension of such challenging and socially impacting professions to be considered in academic teaching.

IV. References

BOQUERA MATARREDONDA, M. (2005). *El lenguaje metafórico de los ingenieros de caminos, canales y puertos*. Valencia: Editorial UPV.
CABALLERO RODRÍGUEZ, R. (2003). "Talking about space: Image metaphor in architectural discourse". *Annual Review of Cognitive Linguistics*, 87–105.
CAMERON, L. & LOW, G.D. (eds.) (1999). *Researching and Applying Metaphor*. Cambridge: Cambridge University Press.
CAMERON, L. (2003). *Metaphor in Educational Discourse*. London: Continuum.
DEIGNAN, A. (2005). *Metaphor and Corpus Linguistics*. Amsterdam & Philadelphia: John Benjamins
GOOSSENS, L. (1995). "Metaphonymy: The Interaction of Metaphor and Metonymy in Figurative Expressions for Linguistic Action". In Goossens, L., Pauwels, P., Rudzka-Ostyn B., Simon-Vandenbergen, A-M.and Vanparys J. (eds.). *By Word of Mouth: Metaphor, metonymy and linguistic action in a cognitive perspective*. Amsterdam: John Benjamins.
JOHNSON, M. (1987). *The body in the mind*. Chicago: University of Chicago Press.
KÖVECSES, Z. (2009). "The effect of context on the use of metaphor in discourse". *Ibérica AELFE*. Vol. 17:11-23.
LAKOFF, G. Y M. JOHNSON (1980). *Metaphors We Live By*. Chicago: The University of Chicago Press.
LAKOFF, G. & JOHNSON, M. (1999). *Philosophy in the Flesh: The Embodied Mind and its Challenge to Western Thought*. Chicago: University of Chicago Press.
LITTLEMORE, J. & LOW, G (2006). *Figurative Thinking and Foreign Language Learning* . Basingstoke / New York : Palgrave Macmillan.
PRAGGLEJAZ GROUP (2007). MIP: A Method for Identifying Metaphorically Used Words in Discourse. Metaphor and Symbol, 22 (1), 1-39.
ROLDÁN-RIEJOS, A. (1995). *"Categorización de términos urbanísticos en inglés y en español: estudio contrastivo"*. Madrid: Universidad Complutense de Madrid [Tesis Doctoral inédita].
—. (1999). *"The applications of cognitive theory to interdisciplinary work in Languages for Specific Purposes"*. En Ibérica (revista de AELFE), 1: 29-37.

ROLDÁN-RIEJOS, A. & P. ÚBEDA-MANSILLA (2006). *"Metaphor use in a specific genre of engineering discourse"*. En The *European Journal of Engineering Education*, 31 (5): 531-541.
ROLDÁN-RIEJOS, A. & J. SANTIAGO LÓPEZ (2009). *"Conceptual barriers in civil engineering language. Implications on gender"*. En *Proceedings of the 37th Annual Conference of the European Society for Engineering Education* (SEFI).
ROLDÁN-RIEJOS, A. (2010). *"The ins and outs of human cognition in the construction of meaning"*m. Available at: http://www.aelfe.org/documents/08_19_Roldan.pdf.
SANTIAGO LÓPEZ, J. (2007). *Análisis socio-cognitivo del discurso de género en el IPA de la arquitectura y de la construcción: propuestas de modificación*. Tesis Doctoral Inédita. Universidad Politécnica de Madrid.
STEEN, G.J. (1999). "From linguistic to conceptual metaphor in five steps" in R.W. Gibbs Jr. & G.J. Steen (eds.), *Metaphor in Cognitive Linguistics*. Amsterdam: John Benjamins: 57-78.
ÚBEDA MANSILLA, P. (2001). *Estudio de un corpus de textos conversacionales en Inglés realizados en estudios de arquitectura: su aplicación al diseño de un programa de Inglés para Arquitectos.* Published by Universidad Complutense de Madrid.

Follow- up practice unit 6

1. Translate the following phrases containing metaphors into your mother tongue:

a. Those referred to human experience/ behaviour

1) The failure of a piece:
2) The damage suffered by the walls:
3) The behaviour of a structure:
4) Deterioration and aging of buildings:
5) Fatigue in a building:
6) Maturity of cement:
7) Curing of concrete:
8) Concrete bleeding:
9) Building pathology:
10) The sick building syndrome:

b. Those based on the human body parts:

1) Concrete pores and cracks:
2) Pillar head:
3) Skeleton of a structure:
4) Concrete skin:
5) Ribs in a vault:

2. Underline the words/ terms that you think have metaphorical potential in the following fragments extracted from different on line journals. Explain your choice briefly.

1) (...) critical building separation (*Strucsaf*-06)
2) (...) the same excitation, acceptable in the case of structures (*Strucsaf*-06)
3) (...) relieving the internal stress that was built (*Compostructures*-04)
4) (...) planning for steel structures subject to fatigue (*Strucsafy*-04)
5) (...) damage level can change structural behaviour (*EngStruct*-05)

2.1 Now, translate these phrases into your mother tongue. Are the metaphors maintained after translating? Mark the metaphorical terms in the new versions.

1. ....

2. ....

3. ....

4. ....

5. ....

3. Using the methodology suggested in Chapter 4 (Corpus linguistics) try to compile a corpus of about 1000 words in the ACE subject matter and identify:

  i. Conceptual metaphor (CM)
 ii. Linguistic metaphor (LM)
iii. Metonymy (M)

Explain briefly the process that you have followed (200 words).

3.1. Extend your search and point out in which genre (s) or part (s) of genre linguistic metaphor has a higher frequency: abstract, discussion, conclusion, report, summary, etc. And in a few words explain their function (descriptive, persuasive, etc.)

3.2. Do the same with metonymy.

3.3. Present 5 examples of architectural or engineering creations with metonymic or metaphorical visual significance briefly explaining your choice.

See: www.LACE.aq.upm.es for sample exercises from this unit.

BIBLIOGRAPHY

BIBER ET AL. (1998). Corpus Linguistics. Investigating Language Structure and Use. Cambridge: CUP.
BOQUERA MATARREDONDA, M. (2005). El lenguaje metafórico de los ingenieros de caminos, canales y puertos. Valencia: Editorial UPV.
BOWKER, L. & J. PEARSON (2002). Working with Specialized Language. A Practical Guide to Using Corpora. Routledge: London & New York
CABALLERO RODRÍGUEZ, R. (2003). "Talking about space: Image metaphor in architectural discourse". Annual Review of Cognitive Linguistics, 87–105.
CAMBRIDGE: C.U.P.
CAMERON, L. & LOW, G.D. (eds.) (1999). Researching and Applying Metaphor. Cambridge: Cambridge University Press.
CAMERON, L. (2003). Metaphor in Educational Discourse. London: Continuum.
CARTER, D. (1994). Introducing Applied Linguistics: an A-Z Guide. London: Penguin.
DEIGNAN, A. (2005). Metaphor and Corpus Linguistics. Amsterdam & Philadelphia: John Benjamins
DUDLEY-EVANS, T. & M.J. ST. JOHN (1998). Developments in English for Specific Purposes. A Multi-disciplinary Approach. Cambridge: C.U.P.
DUDLEY-EVANS, T. (ed.). (1987). "Genre Analysis and ESP". ELR Journal 1.
—. (1988). "One-to-One Supervision of Students Writing MSC or PhD Theses", en A.E. Brooks y P. Grundy (eds). 1988.
—. (1989). "An Outline of the Value of Genre Analyis in LSP Work", LAUREN, C. Special Language: From Human Thinking to Thinking Machines. Londres: Multilingual matters.
—. (2000). "Genre Analysis: A Key to a Theory of ESP?" in Ibérica AELFE, 2, pp. 3-11.
DURÁN, P. & ROLDÁN-RIEJOS, A. (2008). "The Role of Context in the Interpretation of Academic and Professional Communication". In Gibert, T & Alba, L. (eds.) Estudios de Filología Inglesa. Madrid: UNED.

GOOSSENS, L. (1995). "Metaphonymy: The Interaction of Metaphor and Metonymy in Figurative Expressions for Linguistic Action". In Goossens, L., Pauwels, P., Rudzka-Ostyn B., Simon-Vandenbergen, A-M.and Vanparys J. (eds.). By Word of Mouth: Metaphor, metonymy and linguistic action in a cognitive perspective. Amsterdam: John Benjamins.

HERBERT, A. J. (1969). The Structure of Technical English. Harlow: Longman.

HOLMES R. (1997) "Genre analysis and the social sciences: An investigation of the structure of research article discussion section in three disciplines. English for Specific Purposes, 16: 4: 321-337. http://www.aelfe.org/documents/08_19_Roldan.pdf.

HUNSTON, S. (2002). Corpora in Applied Linguistics. Cambridge: CUP.

HYLAND, K (1998) Hedging in Scientific Research Articles. Amsterdam, The Netherlands: John Benjamins.

—. (2004). Genre and Second Language Writing. Ann Arbor, MI: University of Michigan Press.

—. (2008) "Seminar on Genre" at Departamento de Lingüística Aplicada C. y T." Madrid: UPM.

JOHNSON, M. (1987). The Body in the Mind: The Bodily Basis of Meaning, Imagination, and Reason. Chicago: University of Chicago Press.

KENNEDY, C. & R. BOLITHO (1984). English for Specific Purposes. London: MacMillan.

KENNEDY, G. (1998). An Introduction to Corpus Linguistics. Londres y Nueva York: Longman.

KÖVECSES, Z. (2009). "The effect of context on the use of metaphor in discourse". Ibérica AELFE. Vol. 17:11-23.

LAKOFF G & NUÑEZ, R. (2000). Where Mathematics Comes From: How the Embodied Mind Brings Mathematics into Being. New York: Basic Books.

LAKOFF, G. & JOHNSON, M. (1980). Metaphors We Live By. Chicago: The University of Chicago Press.

—. (1999). Philosophy in the Flesh: The Embodied Mind and its Challenge to Western Thought. Chicago: University of Chicago Press.

LAKOFF, G. (1987). Women, Fire, and Dangerous Things: What Categories Reveal About the Mind. Chicago: University of Chicago Press.

LITTLEMORE, J. & LOW, G (2006). Figurative Thinking and Foreign Language Learning . Basingstoke / New York : Palgrave Macmillan.

MCENERGY. T. & A. WILSON (1997). Corpus Linguistics. Edinburgh: Edinburgh University Press.
PRAGGLEJAZ GROUP (2007). MIP: A Method for Identifying Metaphorically Used Words in Discourse. Metaphor and Symbol, 22 (1), 1-39.
RODEAN, L. (1981) "The Passive in Technical and Scientific Writing". In: Lally, T. (ed.) JAC Volume 2, Issue 1/2 .Illinois State University.
ROLDÁN-RIEJOS, A. & J. SANTIAGO LÓPEZ (2009). "Conceptual barriers in civil engineering language. Implications on gender". En Proceedings of the 37th Annual Conference of the European Society for Engineering Education (SEFI).
ROLDÁN-RIEJOS, A. & P. ÚBEDA MANSILLA (2006). "Metaphor use in a specific genre of engineering discourse". En The European Journal of Engineering Education, 31 (5): 531-541.
ROLDÁN-RIEJOS, A. (1995). "Categorización de términos urbanísticos en inglés y en español: estudio contrastivo". Madrid: Universidad Complutense de Madrid [Tesis Doctoral inédita].
—. (1999). "The applications of cognitive theory to interdisciplinary work in Languages for Specific Purposes". En Ibérica (revista de AELFE), 1: 29-37.
—. "The ins and outs of human cognition in the construction of meaning"m. Available at:
SANTIAGO LÓPEZ, J. (2007). Análisis socio-cognitivo del discurso de género en el IPA de la arquitectura y de la construcción: propuestas de modificación. Tesis Doctoral Inédita. Universidad Politécnica de Madrid.
SINCLAIR, J. (1991). Corpus, Concordance, Collocation. Oxford: OUP.
STEEN, G.J. (1999). "From linguistic to conceptual metaphor in five steps" in R.W. Gibbs Jr. & G.J. Steen (eds.), Metaphor in Cognitive Linguistics. Amsterdam: John Benjamins: 57-78.
SWALES, J.M. & C. FEAK (2000). English in Today´s Research World. Michigan: University of Michigan Press.
—. (2010). "From text to task: Putting research on abstracts to work". In Ruiz-Garrido, M; Palmer-Silveira, J.C. and Fortanet-Gómez, I.: English for Professional and Academic Purposes. Amsterdam-New York: Rodopi.
SWALES, J.M. (1985). Episodes in ESP. Oxford: Pergamon Press.
—. (1990). Genre Analysis. English in Academic and Research Settings. Cambridge: C.U.P.
TRIMBLE, L. (1985). English for Science and Technology. A Discourse Approach.

ÚBEDA MASILLA, P. (2001). Estudio de un corpus de textos conversacionales en Inglés realizados en estudios de arquitectura: su aplicación al diseño de un programa de Inglés para Arquitectos. Published by Universidad Complutense de Madrid.

VARTTALA, T. (2001) Hedging in scientifically oriented discourse: exploring variation according to discipline and intended audience. Available at: http://acta.uta.fi/pdf/951-44-5195-3

WIDDOWSON, H. G. (1986). Explorations in Applied Linguistics 2. Oxford: O.U.P.

ABOUT THE AUTHORS

Dra. Ana Roldán-Riejos, ETSI Caminos, Canales y Puertos. C/ Prof. Aranguren, s/n Madrid 28040. E-mail: aroldan@caminos.upm.es

Ana Roldán-Riejos (BA; PhD; Diploma in Translation Studies) is an Associate Professor in UPM Madrid Technical University where she teaches English for Specific Purposes (ESP), in particular academic and professional English for engineers, at the Civil Engineering School.

She has participated in various university-funded research projects on the influence of tutorials on engineering students and also on special integrated actions between UK and Spanish institutions. The most recent one, that she coordinates, deals with a higher education ELP (European Portfolio of Languages) application in ESP contexts.

She has published research articles in international journals and presented papers in conferences on Linguistics and on Education for engineers both in Spain and in many European countries. She has supervised various doctoral theses and belongs to the editorial committee of Ibérica AELFEJournal.

Dr. Joaquín Santiago, EU de Arquitectura Técnica de Madrid. Av. Juan de Herrera, 4, 28040. e-mail: joaquin.santiago@upm.es

Joaquín Santiago López has been lecturing as an ESP expert over the last 11 years in the UPM; more particularly, he has been teaching English for Architecture and Building Construction and English for Academic and Professional Communication to native and non-native English speakers. Additionally his research has been concerned with the development and consolidation of masculinity myths and the influences of gender images and ideologies on the dissemination of socio-cultural models and their building into LACE discourse practices. As a result from this multi-disciplinary perspective, a cross-disciplinary bridge has been established between English for Specific Purposes practice, Cognitive linguistics, gender studies and socio-cultural analysis.

Dra. Paloma Úbeda Mansilla, E.T.S. de Arquitectura de Madrid Avda. Juan de Herrera nº 4, 28040 Madrid. Email paloma.ubeda@upm.es

Paloma Úbeda Mansilla is a Senior Lecturer in the Department of Linguistic Applied to Science and Technology in the Technical University of Madrid. She has been teaching English and Spanish as a second language for 17 years at different educational levels and lecturing ESP for architects since 1994. She taught Spanish in Exeter (U.K) for two years and in several Secondary Schools in Sweden. She has a PhD in Education with a special mention by the Complutense University of Madrid and a Postgraduate Diploma in TEFL from Exeter University. She also lectures Doctorate and Master courses on Language and Communication subjects. Her research interests include teaching methodology, cognitive linguistics and Professional Application in English and Spanish for Architects.